WINTER

Books by Rick Bass

THE DEER PASTURE

WILD TO THE HEART

THE WATCH

OIL NOTES

WINTER

RICK BASS

WINTER
Notes from Montana

Line Drawings
by Elizabeth Hughes

Houghton Mifflin / Seymour Lawrence

BOSTON   NEW YORK   LONDON

For information about permission to reproduce selections
from this book, write to Permissions, Houghton Mifflin
Company, 2 Park Street, Boston, Massachusetts 02108.

Library of Congress Cataloging-in-Publication Data
Bass, Rick, date.
Winter : notes from Montana / Rick Bass.
p.   cm.
ISBN 0-395-51741-9   ISBN 0-395-61150-4 (pbk.)
1. Winter — Montana.   2. Wilderness areas — Montana.
3. Montana — Social life and customs.   I. Title.
F735.B37   1991   90-48703
978.6 — dc20   CIP

Printed in the United States of America

HAD 10 9 8 7 6 5 4 3 2 1

Book design by Robert Overholtzer

ACKNOWLEDGMENTS

Portions of this book previously appeared in *Outside, Antaeus,
The Quarterly,* and *The Boston Review.* Grateful
acknowledgment is made to the editors of these magazines,
and my deepest thanks, again, to LARRY COOPER and
CAMILLE HYKES. Also much appreciated is a Mississippi Arts
Commission creative writing grant, with which I was able to
purchase many of winter's necessities.
Lines from two poems by Raymond Carver, "Looking
for Work" and "Woman Bathing," are from *A New Path
to the Waterfall,* copyright © 1989 by the estate
of Raymond Carver. Used by permission of Atlantic
Monthly Press.
Lines from the song "Night Rider's Lament," by
Michael Burton, are used by permission of Groper
Music. Copyright © 1975 by Groper Music.

For my mother and father

Kent Jingfors, a Swedish muskox biologist, once camped out in the drainage of the Sadlerochit River in Alaska in winter, in an effort to learn how muskoxen survived there. He recalls "days" of brutal cold and darkness when it took nearly all his will to carry out the simple tasks he set for himself. The animals moved slowly through the willows, wary of him. He followed behind with a flashlight, peering closely to see which plants they were browsing. He came to look on them with awe. Anyone who has tried to work effectively in −40°F weather, to contend with darkness in winter for long periods of time or the knife slash of windblown snow at these temperatures, wonders that any creature can endure like this for weeks on end, let alone seem to be at peace.

— Barry Lopez, *Arctic Dreams*

WINTER

Prologue

I'd been in the mountains before. I'd gone to a college built on the side of a mountain, Utah State University, and had never been so happy — not at being young or being in college or being free, but happy just at being on the landscape, moving across such a strange, wonderful land (I'd grown up in Texas and worked, after college, for several years in Mississippi). My girlfriend, Elizabeth, and I went camping out west often. We liked the way the woods there smelled at night and when we first woke up in the morning.

In the horrible, heated summer of 1987 we started our drive toward southern New Mexico, wanting to find the ideal "artist's retreat": a place where Elizabeth could do her painting, and where I could write (separate studios, of course, because we both like to work in the morning); a place near running water, a place with trees, a place with privacy. We wanted a place of ultimate wildness, with that first and last yardstick of privacy: a place where you could walk around naked if you wanted to, a place with a barn, field, and stables for keeping horses, because Elizabeth loved to ride. And if there was an indoor swimming pool and maybe some tennis courts on the premises and a little garden, well, that would be all right too.

Because we were so damn poor, defiantly poor, wondrously poor — but not owing anyone anything, and in the best of health — we were looking for a place to rent rather than purchase, though the thought had crossed our minds that maybe we could buy an acre of land somewhere and build on it. Understand that sharpening a pencil is a great mechanical adventure for both of us; we're a little artsy, I'm sorry to say, but no matter. Such was our fever, the heart's desire, to get out west, to live our lives there. It felt as if too much time had already gone by, we were both twenty-nine, and so we began looking. It didn't matter what state; we just wanted it to be in the West.

We found nothing. Zilch in New Mexico, *nada* in Arizona. We hunted hard in Colorado, but it was crowded — what we call crowded — and all for sale, not rent, or all gone. Maybe a place here and there, but it would be by a lake, not a creek or a river. We took notes.

We scoured Utah south to north. I found two places I liked — one very much, north of Logan — but Elizabeth felt it was a little too close to the dusty shell road, and she was probably right. The forested land in Wyoming wasn't for sale. In Idaho we found one remote place, but we both chickened out. It was treeless, windy, and winter would be coming soon. The idea of building a cabin seemed a little more serious up in Idaho, standing there in that wind, than it had been when we left Mississippi with our two hounds, driving barefoot, with the windows rolled down.

I should mention that it was our practice, in and around towns, to (a) prowl the back roads, and tack notes on the sides of old, fallen-down buildings, asking the owners — next time they came out to chop weeds, perhaps — if we could live there; (b) buy newspapers and read the want ads for rental country places; and (c) visit realtors, who were completely uninterested in renting — interested only in big-money, flash-glitz sales — and who usually showed us the back of their hand. The realtors' piggishness discouraged me so much that as we drifted farther north — July eaten away by the slow miles of our truck, the many nights camped by rushing streams that were paradoxically, wonderfully, not ours, could never be ours — I began to avoid them.

On up toward Missoula, Montana — nothing remote enough — and then farther north, toward Glacier National Park and the little ski town of Whitefish, which we had always liked.

In Whitefish Elizabeth had a feeling, one of those premonitions. We went in to see a realtor on the little main street. It was a slow day, a slow town, and we finally met a realtor, Ross, who was delighted with our plight, and he began telling us about a wild, magical valley up on the Canadian line over near Idaho. Yaak, he said, wasn't really a town — there was no electricity, no phones, no paved roads — but a handful of people lived there year round, sprinkled back in the woods and along the Yaak River. Ross said that it was hard to get to, that you needed four-wheel drive, which we had, and that he had a property for sale there. We pretended to be interested buyers, possibly full of limitless cash. We also intimated that while we preferred to rent, there always remained the option of procuring a certain mysterious investor friend. It's a miracle Ross didn't have us arrested. I was sure, by then, that Ross had come to the accurate conclusion that we were bullshitting him, perhaps from the moment we walked in (hounds chained to the parking meter; bald tires on our sagging truck; or perhaps the giveaway was our sun-faded overalls). Ross suggested, however, that we go look at this strange, far-off little valley, even look at his client's property. A young couple with children were living there, caretaking, for the owner, Ross's client, whose name was Holger. He was in Washington, D.C., doing something vague and undefined for the CIA, an ex–war correspondent who had been shot and jailed numerous times in every unfriendly country there was. For a long time Ross regaled us with Holger's adventures.

Holger used to run a hunting-guide service out of the place Ross was trying to sell for him (Holger had been divorced). Fix Ranch was huge, with numerous guest cabins, a main lodge, hay barns, stables, a greenhouse, a chicken house, extra garages, outbuildings, and a root cellar. It was surrounded by two million acres of national forest, had running water (in the creek as well as in the house, which used a gravity-drain system from a pond up the hill, back in the woods, fed by the spring creek), and a field. The main lodge was three stories tall, each floor lined with big picture windows that looked out at the mountains — Canada to the north, and Idaho just over the next ridge.

It was the end of the world, Ross said. Beautiful, but just too hard to get to. The only way to get in touch with the ranch was by mail or, lacking phones, by short-wave radio — always chancy, and on windy days in spring it was almost impossible for radio waves to get into or out of the valley. There was a small propane generator in the root cellar to provide brief, sputtering electricity — for the electric typewriter and a little record music in the evenings — and there were some twelve-volt deep-cycle RV batteries in a shed, which recharged each time the generator ran; those batteries, along with propane lanterns, provided reading light at night.

We left Whitefish and drove through the afternoon,

seeing no one in the last hour and a half except moose, deer, elk, and grouse, all running across the road in great numbers. White daisies lined the one-lane dirt road.

Heading northwest, we drove without seeing any sign of human life — deeper and deeper into the last and largest spot of unroaded green on our highway map — and we grew heartsick because we were poor.

We kept driving, climbing, and then we came down off the summit and into the little blue valley.

There was nothing but a mercantile and a saloon, one building on either side of the street, and a slow winding river working through the valley (a cow moose and her calf standing in the river behind the mercantile) — and still no sign of life, no people. It was as if they had all been massacred, I thought happily. We knew immediately that this was where we wanted to live, where we had always wanted to live.

We had never felt such magic.

The woman caretaking the lodge answered the door when we arrived — her husband was off in the small town of Libby, forty miles away, looking for work — and she showed us around the ranch, inside and out, with a great, nervous courtesy. She had two children underfoot, and some cats, and looked harried. We moved from room to sun-filled room, still wearing our road overalls and mumbling something about investor backing in New York, down payments, and so on, all to get her to show us the lodge's wonders. Thinking back on it now, I realize that the woman seemed a little frantic, delighted for the company, and often seemed to be addressing some third party, which sometimes appeared to be just in front of us and at other times just behind us. She never seemed to get the perspective of eye contact and spoken sentences down correctly.

In the woods behind the house, deer stood at the edge of the dazzling sunlight, standing back in the dapple of shadows, licking the salt block set out there. The air was hot.

The woman raised chickens, rabbits, her own children, her cats, a wonderful garden, and cultivated a blooming paradise inside a flower- and herb-scented greenhouse. I thought, What a wonderful place that greenhouse would be to write in, and thinking of writing reminded me of being poor again, and it suddenly occurred to me that even though the woman and her husband were living in this beautiful lodge, they were just barely hanging on. Some of the chickens out in the coop were fluttering weakly on their sides in the sun, and it seemed that nothing could be done for them.

"They've been dying all week," she said. Then she turned to us and asked, "If y'all weren't interested in buying it, do you think you might be interested in caretaking? We just decided this morning that we're going to have to move.

There's no work, nothing to do. It would be great for you two — a writer and an artist."

My head felt light and I saw brightness everywhere I looked. Was some sort of trick being played on me? My heart was racing and I could barely speak, and when I did, my voice did not sound like it belonged to me.

"Well," I said, "maybe." I believe that I closed my eyes, trying to draw out the moment, to hold on to the hope of it.

I drove out a month later, driving for days and nights on end in a huge rented moving van that got six miles to the gallon and wouldn't go over thirty-five miles an hour once I crossed the Continental Divide in New Mexico. I listened to country-and-western stations on the radio and talked to the hounds, Homer and Ann. Elizabeth would come out a couple of weeks after me, on the train. I drank black coffee and went through two Ryder trucks — blew a head gasket just outside Vicksburg, a few miles into the journey, and a transmission near Santa Fe — and as I drove, with the windows rolled only halfway down, because the air was already chilly out west, I felt freer and fresher and more daring, more hopeful than I can ever remember feeling.

It was early September and I was driving, literally, to the last road in the United States, a gravel-and-dirt road that paralleled the Canadian border, up in Montana's Purcell Mountains. It was like going into battle, or falling in love, or waking from a wonderful dream, or falling into one: wading into cold water on a fall day. Leaves tumbled down onto those back roads, blew across them. Sometimes, after I stopped for a brief nap, sleeping on the big seat with the dogs curled up on the floorboard, I would wake up and fall-colored leaves would be pasted all over the truck's yellow hood and windshield; and I would almost be shivering.

It was still hot back in Mississippi and in Texas, where I used to live, but it was already cold up in the mountains, up in the North, in this place where I was going to start a new life. The immediate, pressing problem, I realized, was that winter was perhaps a month away. I knew nothing about winter. I had never seen it before, and I felt dizzy with fear, giddy with wonder, anticipating it.

The dogs, I could tell, were worried too, and missed Mississippi. I could tell they thought I was making a mistake.

SEPTEMBER 13

The first overcast day. Wonderful and gray and cold and heavy with moisture when I first got up, but finally, by midmorning, the sun labors, burning a hole through the mist, spreading blue into the valley. I'm eager to see how rain falls here, but know I'll get my chance. It's very quiet and still. You can hear sounds from a long way off: the dogs chewing and gnawing like bears on some deer legs they found near one of the old butchering cabins; birds whose names I don't know, birds back in the woods; a truck passing. Out over the valley it is blue, but up here on the side of Lost Horse Mountain there is a hard silence, and the sun burns through the damp clouds, then is shrouded over. I keep forgetting that I am to eat spaghetti with Dave Pruder and his girlfriend tonight.

When it starts to snow, I'm going to write in the greenhouse. There's a wood stove, and I'll keep warm that way. Two sticks of wood should equal one page, if I'm lucky. I've built a little gym up in the loft of the greenhouse, and will be able to do dips, curls, bench presses, and tricep extensions in a jungle of green, sweating while I look out at winter, when it gets here.

I bought a bottle of wine to take over to Dave Pruder's this evening. Dave is tall and friendly, a horseman, and owns the two horses in Holger's (our) pasture, in addition to the Dirty Shame Saloon. His girlfriend, Suzie, used to be a card dealer in Las Vegas. She's small, red-haired, high-cheek-boned, beautiful. I hope that she and Elizabeth will get to be good friends. Dave is boyish, and his eyes have lots of excitement. He's been up here eight years, and you could watch him and believe, easily, that each year is his first.

September 15

A writer in a valley of workers. Perhaps the novelty of it will allow them to tolerate me. No one asks if I'm going to be staying for the winter; instead, the way they phrase it is, "So, are you going to try to winter here?"

Then I will be a resident — if I last the season. After I've wintered, I will be able to move around the valley with more ease. I will see more things, hear more things, more things will be told to me.

Truman Zinn, Dave's friend, was saying how nice winter is. "A truck might go by your house only once a week, and it will be a real treat. You'll hear it coming from a long way off."

That's what they talk about most when they talk about winter: the silence. Jim Kelly, a retired forest cruiser — wild-looking, handsome, long straight black hair, Indian cheekbones, dark river-stone eyes — says he gets so crazy in winter that he just throws on his snowshoes and runs into the woods, up Hensley Face Mountain and back, a ten-mile trip, just for something to do. He says mountain lions follow him. He crosses their tracks in the snow sometimes, deep in the woods. On his return trip home, he finds their tracks on top of his, shadowing him.

Later on today I am going to bait Homer by saying, "Where is Elizabeth?" He hasn't heard me say her name in over two weeks. It will be interesting to see if he remembers, and then, too, at the train station.

I have a two-week beard. The train station will be fun.

Can't get enough wood! Ran back to the house to use the bathroom — cold and windy, blustery and damp — and when I went out to the greenhouse again, the aspen logs smelled good, popping in the wood stove: good warm dry heat on a nasty day. I'm lucky.

I hope Elizabeth likes cabbage. It's all over the place outside in the garden. I see it when I'm sitting at the desk;

looking out my window, which still has beads of rain on it
from last night. The lettuce stands bright and green from
the rain too. I'm not much on vegetables. I should open the
garden gate and let all the deer in, a Christmas surprise in
mid-September.

One old woman, whom everyone calls Grandma, has lived
up here all her life. Eighty years in the Yaak. Think of all
the things she has missed. But think of all the things she has
seen that the rest of the world has missed. No one can get it
all, no matter where they are.

I don't know how to write about this country in an orderly
fashion, because I'm just finding out about it. If a path de-
velops, I'll be glad to see it — as with math, chemistry, ge-
netics, and electricity, things with rules and borders — but
for now it is all loose events, great mystery, random lives.

If I see a good piece of wood by the side of the road, I
stop and pick it up.

I'm waiting for the propane truck to drive up. I'm going
to buy 900 gallons of propane for the stove and generator
(69.9 cents per gallon). For the family that lived here before
me — husband, wife, two babies — that would have been
three or four months' worth. I am going to make it last five
or six months.

Later in the afternoon. I just returned from Yaak, where I
made a call at one of the twin pay phones, talked to Dick
McGary for quite a while (it was warm inside his store), and
bought ten chances for the Yaak Volunteer Ambulance's
quilt raffle. The quilt hangs from the rafters in the store,
scorching white in the sun that comes through the high
windows. A dollar a raffle, and the quilt is the color of snow,
of winter.

I went and picked up a truckload of wood: such good wood. Thick stumps, already cut, and long lodgepole pieces, dozed up into towering slash piles like toothpicks, where the logging has gone on — clearcuts, usually. Too damn much logging, slash piles everywhere, and wood going to waste, wood that was too small to use but got in the way. Here and there in the slash piles I'll find some larch, which is the driest of woods up here and burns hot; it's everyone's favorite.

Didn't see anyone up on Hensley Face. Didn't want to see anyone. The dogs played on top of the slash pile, leaping from lodgepole to lodgepole, burrowing down into the piles after the scent of who knows what. Their fur is getting thicker.

Driving back afterward with a beer from the Dirty Shame, I met Truman and Dave, who were fixing to go into town proper, Libby. They asked if I needed anything; this made me feel good. When I crossed the south fork of the Yaak and saw our pasture, Dave's old line-backed dun,

Buck, was lying curled up, resting, and the sorrel, Fuel, was standing with his hind legs tucked slightly.

They say Tom and Nancy Orr's cabin has polished elk-horn handles on the drawers, cabinets, and doors. Nancy cross-country skis all winter long, and finds the antlers in the woods.

The wind is lifting and creaking the shingles and rafters over my head. Inside the greenhouse it smells like smoke, like meat, along with the heroic, leftover, holdout smell of spring, of things still growing.

Homer, crazy Homer, is eating tomatoes off the bush, standing knee-deep in flowers to reach them. Ann is napping in a patch of sun. My clothes smell like wood smoke.

Dick McGary said they had done an estimate and found out that it would take $80,000 to run an electrical line out here. Good, I said. Wonderful.

This light, this bronzed tintype light, the light of late summer, Indian summer — it seems to be saying something. This light is everywhere. It's not wavering, not brightening, nor is it dulling: it's just shiny, frozen, existing in itself.

Thinking now about the old woman who's lived here eighty years.

More talk about winter. McGary says they got a foot and a half of snow last Halloween night.

SEPTEMBER 17

Nancy Orr, who came to dinner over at the Dirty Shame, was telling us about her dream hoops. She weaves wreaths out of cane and then braids them with feathers from all the birds in the valley. She covers the entire wreath, leaving a small opening in the center. According to Indian lore, you hang the dream hoop over your head at night, and as you sleep, any nightmares and fears that come to you will get tangled up in the feathers. Only good dreams are able to pass, through the small hole in the center.

It's so spooky, and so sure of its own logic — so fantastic — that you know it must work. Dave and Suzie, as well as Nancy, swear that it does.

A good dinner. All the spaghetti in the world, and beyond. Long, lazy days. Fish jumping out on the river across the road from the saloon. Elizabeth arrives at the train station tomorrow.

SEPTEMBER 19

I picked Elizabeth up at the train station before daylight yesterday morning. Everybody should get to do this at least once: wait like that, in the darkness. I could hear geese flying over the mountains, going south. It's a wonderful thing, just waiting, with the train running late — it was due in at four-thirty A.M. — but knowing that it is coming, and then seeing the light.

All the rest of our old life fell away into the past when I saw that light far down the tracks. The station in Libby is an old one. No one else was waiting for the train at that hour of the morning, and nobody else got off the train, just Elizabeth.

I feel fractured, filled with luck. I gathered her up — she was grinning as if she were fifteen, sixteen years old — and we put the bags in the truck, and she hugged the dogs too.

We stopped and got coffee at the all-night café and then started up the narrow cliff road to Yaak Valley. It was intoxicating to have nothing behind us anymore, and to have everything ahead of us. Deer ran back and forth across the road, passing through our headlights. A month is a long time when your life is new.

We reached the ranch, drove up the long drive, went inside — the smell of logs — and went upstairs. It was still dark and cold in the cabin, cold outside.

When we woke up, the sky was blue and Elizabeth could see that the leaves had started changing already. The wind was still blowing, and it was from the north.

September 20

We got four loads of wood yesterday, about one cord. It's hard getting used to talking and listening to someone else. I'm delighted to have her here, but am surprised at the adjustment I need to make.

We've been having hard frosts every night, twenty-five, twenty-six degrees; but we're in shirtsleeves and barefoot again during the day.

The dogs keep finding more bones. Things seem to die all the time around here.

Things to do: Clean chimney trap. Clean chicken coop and convert it to winter dog residence. Shovel ashes out of fireplace, wood stoves. Git wood.

Good things to know if you are a logger: wear high boots, of course, to protect your ankles against spike limbs, but also wear high-cuffed pants that are open around the cuffs, not unlike bell-bottoms. You want them stopping up high, around the ankle, so that when you are climbing around in a pile of slash, every little branch and limb doesn't go up your leg and trip you. You want baggy, flaring pants, so that if a branch does run up your cuff, then it won't wedge in there and trip you; you can just step out of it. Truman told me this over beers a few nights ago, and I'd forgotten it until yesterday, when I fell. Everyone's so helpful.

I introduced Elizabeth to the McGarys yesterday, up at the mercantile. Dick was talking about how wonderful winter can be; how it got down to eighty below (windchill) their first year up here; how on a cold, clear day in the heart of winter, with no one around, you can look up and see the sky swirling and sparkling with flashing ice crystals — above you the whole sky, crystals falling out of blue air, even though there is no wind.

Check the antifreeze. This is a simple, stupid thing that someone from easy times, from the warm, simple growth of Mississippi, might easily forget.

Everyone back east wants me to send them pictures, but very few of them sound serious about wanting to visit. This is fine with me. I will send them pictures.

A few nights ago, while getting wood, I saw some grouse. They are big, muscular, quail-chicken birds, runners and scooters, and they taste wonderful.

Haven't seen Tom and Nancy since Elizabeth got in. I think about Nancy a good bit. I want to ask both of them questions all day long, all winter — about dream hoops, about ravens, about trapping.

The consensus, unanimously so, is that this is going to be a fierce winter: fuzzy deer already, men's beards growing faster, old people feeling it in their bones, their hearts; the way stars flash and glimmer at night; the way trees stand dark against the sky. Driving back from the Shame late at night, I've already seen snowshoe hares turned completely white. They've staked their lives on the fact, or feeling, that it's going to be an early winter, and a hard one.

Great huge fat white rabbits, like magicians, rabbits the size of large cats, hide out in the darkness of the woods, waiting for the snow that will save them. Evolving all these thousands of years, tens of thousands: the foolish ones were long ago weeded out, the willy-nilly, turn-white-for-no-reason rabbits; those that turned white when there really wasn't any snow coming, or those that turned white too early, had nowhere to hide, and so were quickly consumed, visible prey to wolves, coyotes, hawks, owls, bobcats, lions, everything. And the hares that stayed brown, the ones that did not feel the hard winter coming and did not prepare for it, they got theirs.

So I figure these rabbits know. What a remarkable thing, to bet your life each year, twice a year actually, because they must know when to turn brown again. This year I have been seeing rabbits come out of hiding after the sun goes down, white rabbits hopping across the logging roads as I come down off the mountain with a load of wood — trying, with the windows rolled down, to listen and feel for myself, and to learn rather than always having to be told. I think that I can learn.

There are cars and trucks parked outside the Dirty Shame when I go past — mostly trucks — and it looks warm and inviting, a glow in the night woods. But I've got my window rolled down, I can feel it, feel what I *should* be feeling. What it tells me is that I have gotten up here late to this little valley, maybe too late; but once I'm home, unloading the wood from the truck, smelling the fresh cut of it and feeling the silence of the woods all around me, what I feel, like the rabbits, is this: surely better late than never. With a flashlight, after eleven o'clock, I go back into the night again, driving past the Dirty Shame once more and up onto the mountain for more wood, trying to beat what, like everyone else up here, I now know, rather than think, is coming.

What I feel is that I had better not stop, that I have lost time to make up for.

Hard dreams, of the line-backed dun across the road, his face iced over with snow and sleet, his back to the wind, staring straight ahead as if seeing something far out in front of him; as if watching, through a looking glass, spring and its thaw, its greenness. Dreams of heavy axes hitting frozen wood, of steaming coffee from a thermos, of sitting inside the Dirty Shame, the whole valley — thirty or forty of us trapped by a blizzard — cheering at the football game on the little TV screen, yelling and making bets as to who will win.

What Edward Hoagland has called the courage of turtles, I can see now as the wisdom of rabbits. There is so much to learn. Everything I have learned so far has been wrong. I'm having to start all over. I want to find Nancy today and ask her one more time about those dream hoops, ask her again how they work, ask if there are life hoops as well, though already I know she will tell me that yes, there are.

It can be so wonderful, finding out you were wrong, that you are ignorant, that you know nothing, not squat. You get to start over. It's like snow falling that first time each year. It doesn't make any sound, but it's the strongest force you know of. Trees will crack and pop and split open later in the winter. Things opening up, learning. Learning the way it really is.

All through the forest, they say, you can hear the trees on the coldest of nights: cracking and popping like firecrackers, like cannons, like a parade, while rabbits, burrowed in the snow beneath them, sit quietly, warm and white, saved, having learned — having made the right bet.

Nancy, in the Shame, was telling me about how she hurt her wrist skiing one year. It wouldn't heal and it was all drawn up and forever cramping; a guy told her to stop drinking coffee, and she did, and two days later the wrist was supple again, strong.

I had bad dreams and a recurring headache all night, dreams of dying, murder, cancer, fires, traffic jams. I need a dream hoop.

Evening. I'm writing again, and throwing wood into the stove. Little wrist-size pieces, they burn almost as fast as you can look at them. Energy. Can't have too much wood. It makes the valley smell good. Of course, there are only thirty of us in this valley. This is the proper number. The Dirty Thirty. Next March I will be thirty. If everyone in the world burned as much wood as I am going to this winter, the planet would be obscured, one great wood-smoke cloud. I don't know what to think about that. We're all dirty, but we're all sweet!

I recycle my aluminum! I don't litter! I try to pee on the rocks, not on the soil, to keep from killing things with too much nitrogen!

I remember now how my father pronounces the word "moron" when aiming it at me: "mo-ron." Maybe I am a mo-ron for using wood for fuel rather than the similarly priced propane (though I get my wood for free, with the saw, the ax, the biceps, the deltoids).

We all have dirt in us. Wood is better than coal, but not as good as gas.

No, that's hypocritical, rationalizing. Wood is bad, inefficient, dirty, but it smells good. It's fun to chop, and I like to watch the flames, watch the erratic, pulsing heat it gives, and I like the snaps and pops, and when I'm dead and gone, I'll be glad I used it.

"Mo-ron," the children of the centuries after me will cry. But there will be jealousy as well as anger in their cries (and we are all the same, always have been), and there is wood lying all around, wood everywhere, and it is free, and I have a life to live. Me first, it feels like I am saying. It is my turn, and you may not even get yours.

You should hear my father say it — in traffic, or watching a baseball game, when the manager makes a bad move: "mo-ron." It's like a wave at sea, rolling high on the "mo" and cresting, rolling down into the curl and lick and wave of "ron," sliding softly to shore. The word of the nineties, I'm afraid, is environmental mo-ron, mo-ron, mo-ron . . .

I know I should be burning gas, not wood. I know I should.

SEPTEMBER 21

I took Elizabeth to the Dirty Shame last night and introduced her around. (Grammar rules say not to end sentences with prepositions, but "introduced her" is *not* the same as "introduced her around," and this is the Yaak Valley, and I'm tired of rules.) We drank beers and listened to Ron's storytelling (I don't know his last name; he just showed up here a few days ago). Ron has logged in Alaska and Canada, been on sinking fishing boats, and is, for some private reason, on the lam, and naturally, the Yaak is the perfect place for this: silent, empty, and right on the border. He can run to the woods, cross into another country, and be safe at the first sign of trouble. I'm not sure which side is after him. I don't know what it is they say he has done. I know I like him. He's the best storyteller I've ever met.

Dave Pruder's eyes grow large, like a child's, and his words float, ascend, whenever he gets excited. He'll be wiping dishes, drying glasses, wearing his apron, and he'll look at you, drying faster and faster, telling you where to go see the cow moose and her calf (up Pete Creek Road), or where he keeps seeing that big mountain lion (they call them cats up here). It's hard to picture him riding wild horses, breaking saddle broncs, but he used to. He's tall and stoop-shouldered, not heavy at all, and wears a baseball cap all the time. He's about thirty-five but looks as if he might be at home in a college dorm. Those eyes. He gets enthused about anything, even when he's explaining how the various games of luck in the Dirty Shame work.

I've noticed that the Shame has nothing *but* games of chance — never anything involving skill. Perhaps, in such a demanding, unforgiving country — and anywhere else, I suppose — the absence of a thing in the natural landscape creates a need for it in the psychological one. I'm thinking now of Amish farmers, who lead horribly (or wonderfully) strict and severe lives and always dress in the same drab clothes, but also of the startlingly magnificent, flower-colored bursting quilts Amish women make: the bright yellows and pinks, the deep purples, the zigzagging hexagons of the hex quilts. The women dress plainly, but through their fingers, working the needle and thread, comes the other thing, the thing that is in them but nowhere around them, for their church has forbidden it and it does not exist on the outside: excitement.

The game most often played at the Shame is shake-a-day. When you buy a drink, you put a quarter in the pot and get to roll five dice in an effort to get three, four, or five of a kind. You get to roll only once, and you have to buy a drink to get the chance to win. If you roll three of a kind, you win a free drink, whatever you're having. Four of a kind means drinks on the house. Five of a kind and you get the pot, all those quarters, but you split it with whoever's barkeeping at the time you win.

There's about $200 in the pot now (eight hundred shakes!). My first and only shake I got three ones. Bud Light. It was mildly exciting. It was my first time in the bar, however, and did nothing to aid my hopes of immediately becoming inconspicuous, just one of the guys. But it's past. I haven't played since that first roll, for fear that I might win the pot: stranger comes to town, leaves with all the money.

Another betting game is the snow pool, to guess when the first day of snow will be. That's one I *have* entered, because winning it would earn me a little respect, would show that I was lucky enough or sensitive enough about the land, *feeling* it, to know when the first flakes would come down. Not up in the mountains, where it can and does snow in June, July, and August, but in town, in the Yaak Valley, on the main (and only) road, right out in front of the saloon.

I picked September 28. It cost me a dollar to play. There was one more available spot, in late November, so I bought that one too, November 22. We're in a three-year drought, so who knows.

There's a football pool on the Monday night game, a dollar a square, and some sort of crap game that I don't understand right by the front door. And cards, with Suzie dealing, Suzie winning, beating all the bearded lumberjacks who are seated around her table — six, seven at a time. After muscling wood out of the forest all day, after fixing broken chain saws and water pumps, feeding horses, repairing fences, working trap lines, and doing the other daily work of their lives, the men and women of Yaak are ready simply to sit back and remember what luck looks like, feels like, even if it is not theirs.

The winner is always applauded, teased, jeered. And it never gets to anyone's head the way I've seen it happen in other places — in cities, say. Up here, everyone knows full well it is just luck — invisible, insignificant, and meaningless. Of no use.

SEPTEMBER 22

People who live up here: potters, loggers, trappers, guides (hunting and fishing), an ex–rodeo clown, ex–bronc busters, and a carpenter. A writer. A painter.

Elizabeth and me, we plan to stay up here a while.

SEPTEMBER 24

The "town" of Yaak — the mercantile and the Dirty Shame Saloon, plus the two open-to-the-elements pay phones (with a stump beneath each phone, to use as a stool) — is only six miles north of our cabin. Yaak itself is right up on the Canadian line, on the banks of the Yaak River, which is lovely and slow, meandering through willows and meadows. The river often has moose standing in it, though there are few fish. Massive clearcutting up on the north fork of the valley has washed sediment into the river, so that although the water is still clear, a quarter inch of clay lies like a cancer on the river's bottom — out of sight, out of mind — which blocks the trout eggs from getting to the gravel they need to lodge on, to be properly fertilized, and to hatch on.

So the river flows through the town and the valley, beneath the mountains — beneath the highest, most snow-capped peak, Mount Henry — an idyllic and lazy river.

But the fishing stinks. It's almost Paradise up here, but not quite. And maybe, if I understand correctly, that's what's needed in Paradise, to make Paradise *be* Paradise: a flaw. One small thing, one small evil, to define the wonder and richness of everything else: Lucifer, the bright angel, in heaven; the serpent in the Garden of Eden. Not too many people in the valley will believe it's the sediment that's run down the fishing in the river. They don't want to admit that the fishing's gone south compared to what it used to be, or they don't want to vilify the logging industry, which employs, however tenuously and marginally, a few of them. They simply say that the river's fished out, that it needs only a few transplants, a few hatchery trout . . .

Some days when I cross the river, going through town to get up to Hensley to cut slash, or head into town to the mercantile, which stocks the basics — canned goods, bacon, soda pop, salt, batteries, matches — I feel a twinge passing the barren, beautiful water: green meadows all around, and those mountains, and that majestic timber rising above. Deer — does with fawns — run back and forth across the road in front of me. Sometimes, just as I pass over the lovely, near-empty river, I feel like an old man, like these are the best days that are left — and that the deterioration of our bodies, with which all of us must make our peace, may be trying to happen up here too, in this valley I love so much already. It's one of the reasons I'm in a hurry for winter, in a hurry for snow.

I find pitiful the news accounts of cryogenics, of people who freeze themselves, sometimes even while they're still living, in order to avoid succumbing to terminal illnesses. But I understand them too, in addition to feeling sorry for them. Winter slows things down, for a fact: it can bury and protect, as well as freeze and harm.

The road to the outside world — the road to town — doesn't lead north, toward Yaak, but south, to Libby (nice name), with nothing but mountain between the two, and it's a wonderful road that leads there, an adventurous road, like something from an old movie with the word "pirates" or "Shanghai" in the title. It snakes and winds, switches back, loops and rises, curls up toward the summit, roughly following the cut of the south fork of the Yaak River. Then it descends along the Pipe Creek drainage (named for the soapstone that can be found in the woods, which the Kootenai Indians used to make pipes) into the town of Libby, which, like Yaak, sits alongside a river, the wide, fast, trout-happy Kootenai River. Libby's a mill town, population 2,400, and is polarized, as is much of the West, with respect — or disrespect — to the environment.

There's a real grocery store in Libby, with fresh fruit from California, seafood, and different kinds of cheese. It's got lawyers, barbers, a veterinarian, a courthouse, a library, a photocopying machine, hotels, cafés, sporting goods stores, hardware stores, a bookstore, garages, schools, a train station, children, a movie theater called the Dome, a drive-in that's open only in the summer — and the magnificent Cabinet Mountains, on the other side of the river, rising far above the town, bearing the scars of some hideous clearcuts scrolled across the sides of the mountains like murals, but also still containing some true wildlands, high mountain lakes, and glaciers.

Libby's seams are fixing to pop, but it's thirty-seven miles from us, over a barely passable road, so I hope the explosion won't be heard up in Yaak. Noranda Mining Company, a Canadian-owned partnership, is pursuing plans to dig the world's largest silver mine beneath the Cabinet Mountains Wilderness Area, and is requesting permits to exempt itself from various water-quality restrictions and . . .

I am starting to breathe fast. This doesn't belong in a journal of winter, a journal of peace. You may know my thoughts on these matters, and people will believe in what they believe in. It's not as if I'm going to talk anyone into being for or against clearcutting or sloppy mine operations or dam building. (They tried to build a dam on the Kootenai River, below the thunderous Kootenai Falls!) You're on one side or the other, the battle lines have already been drawn, and sides chosen.

The lovely thing about the road to Libby — the road to the big city — is that, so I'm told, it's often impassable in winter. Our valley can become marooned, snowbound, and what is perilous even in good weather — the road-hugging cliffs on one side and the south fork/Pipe Creek gorge on the other — can be terrifying and suicidal (white crosses, marking fatalities, line the road's edges) on icy roads or in snowstorms. In the summer and fall it can take an hour to go from Yaak to Libby — or, more important, from Libby to Yaak — but in winter it can take twice as long, and often it just can't be done.

And so if Libby wants to chase the depressed commodity cycle of boom-and-bust; if it wants its population to triple, however temporarily, as mining estimates predict; if it wants trailer parks, crime, long lines at the P.O., long lines everywhere, well, it's the town's decision, not mine.

But for now Libby's a nice, quiet, small town on the banks of a beautiful river. Sometimes a moose will run right through the middle of town; one year, a bear wandered into the downtown district on a sunny day and began pulling up the flowers in storefront window boxes. If and when the mine goes in and Libby begins to fester and boil, it'll all be over on the other side of the mountain from Yaak. We'll be back up in the woods, Elizabeth and I, up in the snow, unable to see what's going on and trying to put it out of our minds.

There's an all but abandoned tennis center just on the outskirts of Libby, near the Forest Service headquarters — twelve perfect tennis courts in a little wooded park, with the sound of the big river nearby and glimpses through the trees of the rushing water, with ducks and geese circling overhead. No one ever uses the courts. We have them all to ourselves when we go in to play between errands in town — the scent of pines overhead and the snowy Cabinets trying to block the sky above us. The clubhouse stands abandoned, all the windows dusty, the inside vacant. Some rich old guy, an early logger, left the money to build the place; he must have been some kind of tennis fiend and imagined, in his benevolence, that there would be others in the mountains who shared his love for the game, or would, if only enough courts were built, and in a lovely enough setting.

There's no admission fee; there's nobody around to collect it. We just walk through the trees, past the gate, and onto the courts. The dogs chase each other round and round on the empty courts, but respect and stay off ours. There's just the sound of ravens quorking back in the woods, and the late autumn sun, dry and thin, floating down through the pines, with the mountains heavy above us, longing for winter, winter coming; but for now it's Indian summer, and the *thwock* of rackets hitting the ball squarely.

We have stumbled into the pie, Elizabeth and I, finding this valley, this life. We have fallen into heaven.

SEPTEMBER 25

Besides grizzly bears, grouse, moose, mule deer, elk, porcupines, ducks, geese, owls, rabbits, mountain lions, bobcats, black bears, coyotes, gray wolves (a handful), badgers, martens, fishers, and wolverines, we have a waterfall in this valley, up the Fish Lakes trail.

To get there you walk up a narrow canyon, with ferns, wet and dripping, lining both sides of the steep cliffs. Vinal Creek rushes along the trail, and towering over it are huge larches and cedars — what's called old growth, stands of trees that have never been logged, and which are as necessary to the forest's diversity and survival as old people are to a culture, to a civilization. You can walk up the trail to the waterfall in a light rain and not get wet, such is the overstory, far above. Fairy slippers bloom along the trail, and you know you're getting close to the waterfall when the air takes on that negative, compressed, heavier-than-gravity feeling — the entire forest tingling — and soon you hear it.

Then, through the heavy trees, you can see the magic of it: water charging straight over and down the cliff, water pouring from above.

A fine book about the Pacific Northwest is Raymond Carver's *A New Path to the Waterfall*. I don't know if he ever walked up Vinal Creek, though I like to think that he did. In his poem "Looking for Work," he writes:

I have always wanted brook trout
for breakfast.

Suddenly, I find a new path
to the waterfall.

I begin to hurry.
Wake up,

my wife says,
you're dreaming.

But when I try to rise,
the house tilts.

Who's dreaming?
It's noon, she says.

My new shoes wait by the door,
gleaming.

I don't think anyone should ever try again to write about waterfalls. It's exactly that way: you hurry when you see the falls ahead of you; you stare at the pool beneath the falls and think of fish; and the house, or your world, or the routine of your life, does indeed tilt when you try to look away from the waterfall. It's more mesmerizing than fire, than flame.

Beside the waterfall there's a small meadow and, quaintly, a log bench. Wooden footbridges built in the 1930s crisscross the clear creek; the shadows of trout slide backward and forward over the creek's gravel bottom. I sit on the bench and just watch the water crashing — thunder, yes, with mist swirling up in a vortex, mist rising back up the cliff like smoke, as if trying to retrace its fall, trying to get back to the top.

There is thick, spongy, electric-green moss on all of the boulders surrounding the waterfall, around the deep, turbulent pool. At this waterfall, I've noticed, there is one water ouzel, just one, who sits on the mossy boulders and bobs up and down, exactly like — the Romantics, whoever they were, would slay me for thinking this — the movement of an automatic stapler, straight up and down like the little bird in a cuckoo clock, a rhythmic, mechanical curtsy. It's as noticeable a movement in the still, heavy woods, among the swirl and confusion of the waterfall, as a neon billboard. A small brownish bird does what it must do, I suppose, perching on that lime-green moss and dancing, though it seems a little late in the year for courting.

I remember some of the things John Muir said about a water ouzel he was charmed by almost a hundred years ago. In *The Mountains of California* he devoted an entire chapter to his ouzel, illustrating the bird in various poses and actions, and calling him a "singularly joyous and lovable little fellow."

> Find a fall, or cascade, or rushing rapid, anywhere upon a clear stream, and there you will surely find its complementary Ouzel, flitting about in the spray, diving in foaming eddies, whirling like a leaf among beaten foam-bells; ever vigorous and enthusiastic, yet self-contained, and neither seeking nor shunning your company.

Exactly! In a hundred years — in this narrow canyon in extreme northwestern Montana, anyway — ouzels haven't changed. I feel spirited back in time, whisked away into the past. This is why we have come up here, to find an unspoiled ouzel. It's better than reading a novel, this spinning sense of loss, of good loss, of transport to other worlds — and it's certainly better than writing one.

I watch the ouzel so long, and with such pleasure, that I begin to feel guilty. Muir wrote that the ouzel is usually found alone, and that's what this one is. I'm alone too, having left Elizabeth back at the cabin, and am both missing her and exulting in the emptiness, the deep wilderness.

> He is the mountain stream's own darling, the hummingbird of blooming waters, loving rocky ripple-slopes and sheets of foam as a bee loves flowers, as a lark loves sunshine and meadows. Among all the mountain birds, none has cheered me so much in my lonely wanderings, — none so unfailingly.

For more than twenty pages Muir sings of his ouzel; I watch mine for perhaps twenty minutes.

One more thing from Carver, from "Woman Bathing":

> Naches River. Just below the falls.
> Twenty miles from any town. A day
> of dense sunlight
> heavy with the odors of love.
> *How long have we?*

It's late in the day, and I start to think of Elizabeth. I head back.

SEPTEMBER 26

There's an owl living in our field that looks as if he stands four feet tall. We looked him up in a book: he's a great gray owl. He likes to sit on various fenceposts, motionless, watching us as we walk to and from the truck, the woodshed, everywhere. Cocking his head sometimes, he leaps down into the tall grass, or into the hay around the barn, to grab mice, which he then flies around with — the mouse hanging from his beak like a limp rag before landing on another fencepost or a stump. People say they've seen this owl around Fix Ranch forever. I count his misses,

tallying them against his attempts, in his war on the mice, and he usually goes about nine-for-ten. It's a surprise to see him miss.

One morning last week as I was writing in the greenhouse, a coyote came trotting out of the woods, sat by the edge of the trees in a patch of sun, and watched the road. The wind was ruffling her fur — I have no idea why I believe the coyote was female, only that I do — and after a while she got up and slunk into the tall grass. Then she began pouncing, chasing a mouse, trying to pin it with her paws, hopping, lunging. Finally, just when I was ready to believe she was hallucinating, that there was nothing in that tall grass, she plunged her snout down and came up triumphant, a wriggling, writhing mouse in her jaws.

Immediately — whether for play or because the mouse bit her, and it almost seemed that it could have been the latter — she flipped the mouse high in the air, twenty feet or more, and then stood on her hind legs, like a person, waiting for it to come down, and she caught it again. She did this several times. A few minutes later she trotted back to her spot at the edge of the woods and lay down and ate the mouse. Afterward she looked out at the empty gravel road, and at the field beyond that, for a while.

Things were going through her mind; you can't tell me they weren't. She just sat there in the sun and the wind, with her fur ruffling, getting ready for winter.

This field is full of mice. They surround the barn, hiding in the hay, getting ready for winter too. Some of them will never see it. There are other forces out there besides just winter.

But the coyotes and the owls, they are also responsible to the force of winter, to the brunt of its coming power. It exerts a competitive rather than a unifying force on the whole forest. It is dynamic, rather than static.

They catch mice the way I gather wood. We're all close, we're all tied together.

SEPTEMBER 27

The glory species up here — and they're the ones that excite my soul when I walk through the woods; they make the woods be *woods* — are the gray wolf (of which I've seen two) and the grizzly bear (one). But the rarest species, perhaps the rarest in the Lower Forty-eight, is the woodland caribou — a subspecies of the barren ground caribou — the Alaskan caribou that you see on the *National Geographic* specials.

Caribou used to roam the old forests of the northern United States, not just in Montana but in Minnesota and all the way to Maine. Perhaps I'm no better than the so-called yuppies — that hideous word for which, sadly, there is no substitute — in that the way they crave money, possessions, and security, I crave wilderness. I want to hoard nature the way they might hoard sports cars. What I'm saying is that I don't want only gray wolves and grizzlies up here (nearly any other wild valley in the Lower Forty-eight would be thrilled to have either of those two species), but caribou too. There are names on the old maps of this area that break my heart, names like Caribou Creek and Caribou Mountain. The mapmakers didn't give us those names by accident.

But I missed out on it. There aren't any caribou up on Caribou Mountain now. Just ghosts.

Jasper Carlton writes that "caribou have an acute sense of smell, but rather dim eyesight and poor hearing. They are unwary at times and inquisitive about objects they cannot smell. . . .

"Caribou may appear awkward (particularly calves) due to their long stilt-like legs and large, almost pie-shaped hooves. Yet these features enable them to move freely in deep winter snow and in wet, boggy terrain."

They look like elk but they're smaller, and both males and females have antlers, big, rocking-chair antlers that curve back over their bodies and then sweep forward again, out over their heads and beyond, like the antlers of their near-cousins, the European reindeer. What was it like in the 1930s, say, when caribou still moved through this valley — secretive and always in deep woods (although they would approach a stranger when they could not smell him, if the wind was wrong) — just a few rare wandering caribou, a herd down from Alaska, down from Canada? What if you were a child, looking out the window of your cabin? What if it was around Christmastime, at dusk, and you saw that small band of caribou moving through the snow?

One last herd, about twenty animals, still occasionally wanders across the line into our valley, and into northern

Idaho as well. Some of these caribou have radio collars on them, having been trapped by the Idaho Game and Fish Department, though even with the collars it's hard to locate them in the heavy woods. It's possible that the remaining herd's gene pool is so reduced that they're doomed without transplants, reintroductions. But the U.S. Fish and Wildlife Service and the U.S. Forest Service won't even list them as endangered species, for fear, is my guess, of all the restrictions that would then be placed on the few remaining stands of old-growth forest up here, which is exactly what the caribou need to survive — to browse the colorful lichens that grow beneath the towering dark canopy. It's clear to most people that the caribou's time has come and gone, that they've had their heyday, their fun, in the Lower Forty-eight.

Since 1900, Jasper Carlton writes, there have been only 158 reports of woodland caribou in or adjacent to Montana; this makes them more common than sightings of Santa Claus, but much rarer than UFO sightings. The last sighting of a caribou in Yaak Valley was in 1987: a lone bull, part of the Moyie River herd that lives twenty-five to thirty miles north of the Canadian border.

Because woodland caribou's home range can extend beyond thirty miles, it's still conceivable that, if you were in the right place, perhaps looking through a cabin window, you could see a caribou, or maybe even the whole small herd, go walking through the tall trees. There are still lots of trees up here — lots of sprawling clearcuts too — but there aren't many of those medieval forests, the ones the caribou like to wander through, looking for lichens — the old-growth forests that the spotted owl and the loggers have made famous.

As I said, they look a little like elk, caribou do. If the government were to reintroduce a herd up here, they'd need to educate the hunters. But the old-growth forests — the tiny fraction that remains of them — would have to be protected first.

One last herd wandering through the woods, back and forth across the border — ghostly, rare, animals one might expect to see only above the Arctic Circle, animals that live only in the oldest forest a valley has to offer . . .

It's time to stop cutting old-growth forests. We'll cut them all, and then the same cries of more jobs, more money, will go up, and where'll we be then?

I don't mean to rant. I'm trying to keep this polite, low-key, respectful. Quiet. Falling snow. But inside, I rage. Sure, cut lodgepoles selectively. But the big larches, the last giant cedars? When there are so few left, and when they're so important to wild nature?

Some people want money, other people want caribou. You have to draw a line and stand on one side or the other.

Living up here in the woods — just a mile or two from Canada — I feel as if I've got my back up against that line, up against a wall, and, like the caribou, there are increasingly fewer places where I'll fit in.

Perhaps that's what drives so many wilderness advocates — fear, as well as love.

SEPTEMBER 28

I keep singing psalms in this journal. But light isn't anything without darkness out there to define it. And even in these dream-mountains there's darkness, on the perimeter of the valley, hemming it in: the knowledge that the valley can, like everything, change, and be lost. But that turmoil outside the valley — the quickening pace of the outside world — is what helps sharpen the focus on the loveliness we've found here in the interior.

I can picture getting so addicted to this valley, so dependent on it for my peace, that I become a hostage to it. And sometimes, being human, Elizabeth and I have to ask, What are we missing? Usually the easy answer, the quick one, is not a damn thing. But some days — here, as everywhere, I think — a longing sweeps into the valley like a haze. But we can't define it, can't pin it down — and it passes soon enough.

SEPTEMBER 30

The last day of the ninth month and we have a heat wave, eighty degrees. I'm shirtless, perspiring, splitting dry lodgepole stumps with satisfying cracks when Kenny Breitenstein comes up the driveway in his tall red truck. He introduces himself and asks if I want to buy any larch. Larch is good stuff, and I'm interested. Kenny's truck towers above me, and I have to squint, looking up at him; he's far too polite to get out on this, his first visit. The larch he's selling is dry trees he felled last fall when he was building a fence along his property. He's selling it, downed, for $50 per thousand board feet — about $25 a cord. I'd cut and load and haul it myself. It's only a mile or so away.

Larch (also called tamarack) is a huge tree, not unlike cedar and the big redwoods, and it grows to immense diameters, cathedral heights; it's what, more than anything else, gives this end of the Pacific Northwest the flavor of the Pacific Northwest. There are some great ancient larches up behind the cabin, and there are doomed ones too, doomed ten, twenty years ago, lying on their sides on Breitenstein's land half a mile away, deep in his woods. They were felled between the smaller trees, so that they look like giants from another age, which they are: dark, rotting giants spangled with vivid green-velvet moss, resting among springy ferns. The heart of these felled trees is a luxuriant amber-orange, like pumpkin meat. The wood burns hot and good, clean and fast: nothing burns like old dead dry larch. Not even the Forest Service crews know where to find much of it — not the twenty-four-inch-diameter larch of firewood quality, most of it already killed by rot or age or by a great fire in 1931, which burned all the way from Spokane to Kalispell, 250 miles of fire.

There are lacy-needled saplings and young larches that turn an eerie gold in early fall, and then later in the fall the valley is filled with their flying gold needles, which swirl through the air and cover the roads with soft bright padding as the fall winds begin. Sometimes these younger larches grow out of the felled larches themselves, which overambitious loggers cut down and then were unable to skid out. These huge carcasses are like hidden, forgotten treasure, rotting in the silent woods. I see downed larches like this all the time on my walks behind the cabin.

But Breitenstein has this good stuff all over his land, and he is eyeing my shirtlessness in this heat of almost-October, and I can tell he wants snow, wants blizzards, wants wood-buying, wood-selling cold. But he's also a friendly guy, tall, crew cut, with gold teeth everywhere, even across the front, and dark, intelligent almond eyes, and, for some reason, only half a faint mustache — the other side gone, shaved away. Perhaps, I think, standing in the heat, looking up, this is his way to disorient customers, to appear fiercer, to drive a harder bargain. I'm dying to know why he has only half of that mustache instead of all or nothing, but it is not yet time to ask, and I'll just have to bide my time.

I tell him that I'd like to buy some of that wood. My children may never burn larch, may never bury their chain saws or axes into its exciting heartiness, but I'm not going to let it rot, either, and even if it were twice the price I'd buy it just for the chance to handle it, cutting, quartering, loading it into the truck the way they used to do twenty years ago when they thought it would be here forever, when a giant tree was only a challenge to be felled.

It's sick, it's perverted, it's foul, but someday, somewhere, if for some noble, valid reason (though I can't imagine one) a person has to cut down one of the remaining monster larches, I'd like to watch it go. I'd like to see how many saws it took, how many chains snapped, how many times the loggers stepped back and mopped their brows or leaned forward and rested their backs. I'd like to see and hear the end of it, the crack and heart-throat splinter at the base, and then the popping as it fell to earth, taking out all the lesser trees around it with its branches, and the shudder, the jolt of the earth, when the tree landed. I love the larches, love to sit with my back against them when I come across them in the woods, larches either standing or felled. I would never cut down one of the giants myself, but someday, if someone else was felling one, I think I would be compelled to watch, like a hanging in the old West. I do not think I would be able to turn away.

It's almost one o'clock in the morning but things are piling up. I've got to take care of at least a few things or they'll bury me, they'll just keep adding up, coming at me. Holger, the ranch's owner (he now lives in Florida, retired, and catches tarpon in the Gulf Stream, putters around in an old boat, fishes barefoot), says to put sixteen-inch tires on my truck instead of the standard fifteen-inch ones. He says that will raise the truck off the snow about four inches or so. It's a thing I'd never have thought of, not in this lifetime, anyway. A beer for Holger when he comes out on October 25, for the opening of elk season.

My beard has stalled, gone wild; it looks hideous. I was writing a check for a gym membership in Libby ($20 a month, weights and shower room), and the girl was eyeing my check, noting that it was a new account, asking for a driver's license, and she wanted to know if I'd found a job yet. I didn't see that it was any of her business, and told her so; I wasn't looking for any jobs. But later, looking in the locker room mirror, I saw that my eyes were bloodshot — I'd been driving all day — I felt contrite, and on my way out I apologized. She was surprisingly cheerful, utterly gracious, and accepted the apology.

More things, but I'm winding down. The weatherman on the radio at the gym this evening said there was a "zero percent chance of rain." I'd never heard that before: zero percent. I liked it, that certainty. No namby-pamby hedging, no "almost nil," none of that "ten percent" stuff like I used to get in Mississippi, even in the middle of a drought. He didn't think it was going to rain tonight, that radio weatherman, and he said so.

The deer have melted back into the woods, have stopped coming down to the salt lick in the back yard. Holger says they can tell when hunting season is coming. He says they'll be back in the winter, though, when the heavy snow is down. They'll be right outside the window looking in, steaming up the frosted windows with their breath, trying

to get to where it's warm; they'll break into the garage, into the barn, and eat the hay.

I got an unnerving letter, out of the blue, from my friend Jamie in New York. His girlfriend has a cousin in Alberta, Canada, who says that it gets so cold up there on their cattle ranch that the bulls' ears and testicles freeze and fall off, and Jamie promised to send me a set of earmuffs. A very funny fellow.

My eyes are sandy, tired. The dogs long ago sacked out. I still have so much wood to get. I need an electric winch for the truck's front bumper, to pull myself out of the snow when I slide off the road. Cross-country skis. A new transmission for the truck. Snow chains for the tires. Snowshoes. Snow shovel. Etc. Etc.

I got the radio phone today. We drove over to Sandpoint, Idaho, and paid $550 for a used sixty-watt radio, the most power the FCC will allow for homes. A man named Emmett Johnson is coming up Monday to install it. He's a bluestreak curser in his sixties, an electrical wizard, a genius. I get dizzy with awe watching him work, listening to him explain electricity to a dolt like myself, explaining it in the simplest of terms, baby talk, but still I don't understand. He's an ex–Navy man. You should see his house: wires slung from the ceiling like spider webs, electrical strands everywhere, radios half-opened on desks with their guts spilled out, nuts, bolts, transistors, cable coils, devilish-looking tools . . . He installed a CB radio in my truck, and I learned some good new curses, and a little bit about electricity — enough to know I should go back to Emmett, for sure, to get him to install this radio phone.

He's brim full of common sense, running over with it, as much of it as anyone I know. He's like my father and my grandfather: he makes things work, no matter what the obstacles.

I can now sit in my cabin in Yaak and talk or listen — through the static of wind, of aspen leaves blowing through the woods — to Tokyo via a dispatcher in Libby, forty miles south. Not my idea of fun, having such a machine in the cabin, but it's a necessity. Elizabeth's mom, and my parents too, are pretty riled about not being able to get in touch. And business, business: always, you can say something to someone over the phone about your business; there's always something to say, and sometimes a need for it.

But there aren't going to be any pizzas delivered up here. And I'm not going to leave the radio on all the time. It plugs in to a twelve-volt battery line, but most of the time there'll just be the nothing sound of snowfall, split-shattering silence — a few tree limbs cracking, maybe, a little wind stirring. I will keep the phone unplugged most of the time,

and use it only to call out. The radio phone is necessary and expensive as hell, but a black sheep.

Time to swat a slow, late-night, leftover-from-the-day fly.

Time to go to bed.

O CTOBER 1

A logging truck is squeaking and rattling down the road, a force, the sound of a howling wind. A hearse for trees. There's never enough of anything. Everyone thinks this country will never run out of timber — of lodgepole, fir, and spruce — but if the giant larches could disappear, then so too can the others. I'm worried about electricity coming to this place someday.

Today's schedule: type, cut wood, split, and stack. The greenhouse has half a cord and needs eight. The woodshed has eight and needs twenty. The kitchen pile, on the side of the garage — wrist-size pieces for cooking — has two cords and needs ten. That's the toughest chore, splitting wood for the cook stove — splitting and splitting and splitting again, taking a fine fireplace-size creaking yellow heart-log of wood, a barrel-size piece, and essentially making it disappear, knocking it down with the maul into toothpicks, which will be gone in a flash, consumed by a quick burst of flame. But they'll cook the beans, those little sticks, cook the bacon, boil the water for coffee, and it needs doing.

Elizabeth is baking bread this morning. She's got a good Mormon recipe that involves punching the dough repeatedly. Every time it rises, you're supposed to punch it down with your fist. A woman from Salt Lake City gave the recipe to her, and that's the word she used, punch, and she showed Elizabeth how to do it on an imaginary lump of dough: jab, jab, jab with her fist balled up, her face scrunched and squinting.

So, Elizabeth is punching dough today. I'll be splitting
wood. The dogs will be chasing chipmunks in the wood-
piles. Something else will be happening out there in Chi-
cago and Houston and Tampa and Philadelphia and even
in New York. Above us, thirty thousand feet aloft, there'll
be air masses moving, weather patterns developing; beyond that, going into the blackness, out to the cold stars,
there'll be who knows what going on — comets, milky
things, maybe strange sounds, orbits, and gravitational
pulls — but down here on the Fix Ranch we will be splitting
wood and punching bread. There's a rooster crowing down
at Terlinde's place, almost a mile away. It's ten o'clock in the
morning.

Splitting logs with the maul, I was dressed for safety, wearing boots, goggles, and ankle guards for when the quarter-
and half-pieces kick back and hit me occasionally. I hadn't
thought about the drawstring of the sweatshirt I was wearing, however, and when I was raising the maul over my
shoulders for the swing, I got the drawstring wrapped
around my wrist, which distracted me on the downswing,
and I almost missed the log and cut my foot off.

I promptly took the baggy sweatshirt off and split wood wearing just my undershirt. I need to know these things.

OCTOBER 3

The first truly cold day was two days ago — all of the nights we've been here have been cold, some more so than others — but then it was warm again yesterday, though we've been getting some high winds all during the day, even in the mornings. The moon is nearly full, bright and blue, a great aluminum coin; as large as it is, it looks out of place in these mountains, fighting its way up over the ridgetops, trying to shine through the trees around eight or nine o'clock each night. A full moon over the prairie, or in the city, or even over a mountain lake, looks full and just, and things are right. Night is king.

But I'm not so sure out here. Again, I cannot explain it — and it's beautiful; what full moon in the woods isn't? — and perhaps I'm just a little threatened by it because it's so damn close out here. You can see lunar craters and seas in detail, unbelievable detail — hell, I'm a couple thousand miles closer to it, and in cleaner air, than I was before; it *ought* to have more visible detail. It comes up so low over the ridgetops that it seems to be stalking me, or at least watching me; and sometimes, depending on my position — driving down from Idaho, for example, back toward Libby and the river bottom — I'll be above the moon, looking down on it.

So. Perhaps what I mean is I've gotten used to the nights being so dark here, so many stars, such privacy. And then this great blue floodlight of a moon pulls into town. No wonder the dogs barked at it that first night:

a ferocious, hackle-raised, head-up-through-the-tall-grass woofing charge, running up into the woods to go bark at it.

We had a strawberry break yesterday, as I split wood. Elizabeth brought big red ones from the garden out to me on a plate with a little powdered sugar on them. It was a cold day but I was shirtless and sweating.

A couple of evenings ago — that cold day — Elizabeth and I took the Falcon and the dogs up to my wood-cutting area on Hensley Face Mountain. There was a wonderful larch tree felled, thick and bright orange, and I had cut a couple of big logs off the end of it, and wanted to split them and load them in the back of the Falcon. (The truck is in the shop awaiting a new transmission, which had to be flown in from Japan. Could be two days, two weeks, or longer. The plane could go down in the sea, as they are doing these days, with my shiny new transmission — brightly oiled steel, slick bearings, grease-packed teeth, $1,600 — sinking and landing on the ocean floor, kicking up a *poof* of silt. There it will lie, through the centuries, while I winter in Yaak in a little Ford Falcon, with the back seat pulled out to make room for wood and fourteen-inch street tires and no heater, no defroster, nothing but a good strong true en-

gine — for now — with an eighth of a million miles on it. It has seen the whole country over and over . . .)

Anyway, we got the larch split and loaded, with dogs, firewood, Elizabeth, and I crowded into this tiny old rumbling car — and the engine barely turned over. It sounded like the battery was almost dead.

So instead of stopping on the way down and getting out to go look for grouse, as we'd planned, we headed on back into Yaak; I had to make some more calls from the town's erratic pay phone. Not thinking, I shut the engine off, went into the phone booth, and made my calls. Of course, when I went back to the car and tried to start it, nothing: just that rapid whir-clicking.

How old do you have to be before you learn not to drive old cars, that they just aren't made to be driven? That they're like people, and after a certain age they should be retired, not entered in log-pulling contests, snow-sledding gauntlets, cross-country marathons, and the like. That old cars should not be driven any more than sixty-year-old men should play football, or eighty-year-old women should sprint, play tennis, and backpack.

But maybe that's why I do it, regardless. Maybe that's why I stay with old cars forever, and why everyone in the valley does. In fact, my '84 Nissan truck (the one in the shop) is far and away the newest and brightest thing on the road up here — *not* on the road, that is.

Breitenstein drives a '59 Dodge, a big, red, army-looking thing. The mail car is an old Subaru with a tall rat-tail CB antenna sprouting out of the center of its roof. Look out for the mail car. Don't hit the only link with the outside world, whatever you do.

Dave Pruder has a maroon-and-white '76 Chevy Blazer. Blazers were new and socially desirable when I was in high school. The rich kids used to rumble into the parking lot in them. Dave's Blazer isn't something any rich kid would be seen in now, though. It's crumpled, but still runs.

Jimmy Marten's jeep could qualify as wedding material, because it's both old and vaguely blue. And so on and so forth. Just old cars and trucks, running up and down the south fork road (from Libby to Yaak), running up and down Yaak River Road (from Troy to Yaak). No one ever uses the remaining road, the one leading out of Yaak and over the mountains and across the lake into Eureka, because, quite simply, quite beautifully, the road's just too bad to be used for any reason other than a picnic.

Deer walk down the center of this unused road, paved long ago, as if it were a weedy, overgrown logging road, a skidder trail cut through the woods. There are switchbacks all the way down and up and then down again. No one from Yaak ever has business in Eureka, a nice little town itself — but with electricity, we sneer. If Yaak folks absolutely have to have something, they go to Libby, which is twenty miles closer, and bigger by a few hundred people. Moose lumber, too, across this Yaak-to-Eureka road, and parts of it are caving in, falling off the high line of the cliff it follows for so much of the way. I love the road.

So the old cars and trucks run back and forth on our two roads, monitoring the CB, scooting out of the way of logging trucks, pulling off and hiding from them when we hear they're coming. Like a crazy man's game of hide-and-seek — or even worse, red light–green light — the logging trucks broadcast, every mile, their thunderous location, barreling upstream or downstream on the little one-lane roads. And it's something to do; it makes the drive, even into Yaak, anything but boring. You've got to have a CB radio. Otherwise there's no telling what's going to be coming around the next curve, with no room to pull over. You might end up in the gorge.

But I digress. I still have an old Ford sitting by the pay phone, loaded down with wood and dogs and girlfriend. The chain saw in the trunk is worth perhaps twice what the car is worth. I didn't have any tools with me; they were (and are) all in the Nissan.

I was embarrassed to be more or less helpless, and yet I knew what the problem was: the mo-rons at the backwoods garage in Mississippi. The utter dolts (the reason I used them was because they were cheap) put the wrong size belt on when they rebuilt my old alternator. (They are dolts, but at least they know how to rebuild alternators, instead of going out and buying a new one, instead of having to depend on someone else; and that makes me more of a dolt, because I do not yet know how to do that.) This belt was slipping, wasn't turning, wasn't charging the battery (as was its duty) whenever the engine was running. I could have tightened the belt somewhat; I could have loosened the alternator mount and slid the alternator farther back, using a broomstick for a pry bar; but I didn't even have a pair of pliers with me.

The simplest solution was just to get a jump and limp on home the six miles to the cabin, where I had a battery charger and a damn pair of pliers, at least. Then I could jump it again, if necessary, and limp on into Libby like the most Beverly of Hillbillies — shocks squeaking, loose fan belt whining, dogs fighting in the back seat. But the key, of course, was getting back to the cabin.

I was glad I was in the town of Yaak (mercantile and saloon). I was glad it was not winter and I wasn't two miles up Hensley Face and night wasn't falling (which it was) along with the temperature. All I had to do was walk across the street into the bar and ask for a jump — but damned if I was going to do that, either. Too much evil pride: the newcomer.

I sat there, with night falling, and let the engine cool. After a while I tried it again and it turned over, just barely. I got lucky.

We drove home into the moon. It was cold. We don't have enough wood in.

OCTOBER 4

The landlord, Holger, is a nice guy — terribly interesting, with all his tales of foreign prisons and espionage — but he's still, albeit at long distance, a landlord.

Several people up here caretake ranches and cabins while the owners, the lords, sit out the winters (and, as in our fairy-tale case, the years). Tim Linehan, who was an all-American soccer player in New Hampshire and who's a rabid fly-fisherman, caretakes an abandoned lodge on the shores of Vinal Lake. Mike Canavan, who used to play linebacker for Syracuse, cruises timber for the Forest Service, shoes and trims horses, and caretakes a ranch on Obermayer Lake. Even Breitenstein watches after property for some people who own a cabin on Lake Renée — a big place, even farther into the wilderness than ours.

Breitenstein, like an executioner, stalks the woods around Lake Renée with his chain saw, cutting out any trees that get infested with pine-bark beetles. He fells the trees and leaves them to rot so the beetles' eggs don't hatch, or he skids the logs, often with a horse, out to the road and then cuts them up for firewood. Breitenstein the executioner, keeping the woods lovely and clean, green and growing — neat, orderly, medieval.

Mike Canavan, in addition to his other jobs, also guides hunters in the fall. He knows the valley well and, as a result of his cruising job, always has stories about sightings of wolves, bears, and mountain lions. He came out one day to help take care of Dave Pruder's horses, Buck and Fuel — to trim their hooves, I think, or maybe float their teeth, something technical — and corralled the horses, but then he couldn't halter Fuel.

Fuel spooked and ran right at Mike, but instead of dodging Fuel, Mike hit him head-on, high in the chest, tackling him. He reached up and grabbed Fuel around the neck, as if bulldogging him, and then got dragged around the corral for a while — never letting go, lifting his feet to keep from being trampled — until Fuel finally calmed down, or maybe was worn down.

I was standing on the other side of the corral, the safe side, watching Mike.

"The halter, please," he said through gritted teeth, his arms still wrapped around Fuel's neck.

I trotted out and handed Mike the halter. He held on to Fuel's nose with one hand, pulled Fuel's ear down and bit it, and slipped the halter on with his free hand. Then Mike straightened up and led the horse over to the hitching post. Neither seems injured.

Mike has a story about everything. He has a bunch of Sasquatch stories, though he's never seen one. The most convincing story is about the trailer down on Yaak River Road, where an old man — I forget his name — lived with his dogs. It was a rickety mobile home, up on cinder blocks and stumps, out in the woods, as everything is.

One night the dogs began barking like crazy. It was real late. The door was locked, but the old guy heard something trying to get in. The doorknob was rattling, and the dogs were scared, growling and hiding behind the refrigerator — not aggressive, the way they usually got when they heard or smelled a bear.

Mike swears this is true. He gets offended, which is a rare thing to see, if you doubt the veracity of the story. If you do, he'll drop it entirely, with a shrug and a who-cares, your-loss attitude.

The rest of the story has Sasquatch, Bigfoot, or whatever the hell it was, pulling the doorknob completely off the little aluminum trailer and then leaving. Perhaps it was simply seeking shelter or had seen the old man going in and out that door, and was trying to figure out how it was done. Mike's seen that torn-up, knobless door and says he doesn't know whether it was a huge bear or Sasquatch. The tracks were like a bear's but were far too big.

There are other Sasquatch stories, a startling number of them, but my favorite Canavan story is about a landlord, the owner of a cabin Mike used to caretake.

Since there isn't much livestock in this valley — a few horses, a handful of milk cows, maybe a goat or two, and some rabbits — most caretakers' jobs involve nothing more than keeping the pipes from freezing in winter — though it's not that easy a task.

When the wind comes down off Hensley Face Mountain some years, beating against the walls and windows and pressing against the base of the cabins, the temperature dives to forty below, with a windchill of seventy-five or eighty below, and it's impossible to keep the cabin warm enough to prevent the lines from freezing. The trick is to shut the valve off at the creek, so the pipes in the cabin can be drained completely in anticipation of a quick and strong storm. But often the storm comes in too fast (there's no radio to forewarn, certainly no television, across most of the valley, just word of mouth among hermits) — or you'll be out hunting or cutting wood — and it'll be too late, your pipes will freeze, and freeze hard.

The cabin Mike used to caretake was underinsulated, and the landlord was not a particularly pleasant man, so their relationship was strained beyond the usual tension between landlord and tenant: the owner was slaving away in California while the caretaker, the ruffian, was enjoying the fruits of the landlord's labor.

In only a few seasons, the water had frozen on Mike several times already, and each time the cabin required repairs — new copper tubing, new welding, and so on — and Mike was about half tired, as he put it, of "dealing with the sonofabitch in the first place."

One day the pipes froze again, solid, worse than ever before. Mike had dutifully called the owner to inform him of this change in status. The owner was going to be driving out for a cross-country skiing vacation, or some damn thing, but it was such a long and hard freeze, so bitter, that Mike couldn't get the pipes thawed out. Often a cabin will get so cold and locked up that there's nothing to do but wait it out, wait for spring. Mike and the owner were both afraid that's what had happened.

Mike kept fires going in the fireplace and the wood stove anyway, roaring hot fires stuffed with larch, in an attempt to warm the cabin enough to get the water to flow again. But the pipes were probably frozen all the way back to the source, back to the pond on top of the hill, for all Mike knew — frozen underground, even. Still, he kept shoveling wood into the stove, doing the only thing he could do: feeding the fires.

He's not exactly sure how it happened — a stray spark rising to the roof, perhaps — but the cabin burned down one day while Mike was out cruising timber. Mike swears that this is the absolute truth. I think about him swinging from Fuel's neck, being carried like a flyweight all around the corral, getting bashed against the fence but not letting go, and I believe him.

"I called him up," Mike says, "and I told him I had some good news and some bad news." It's at dinner that Mike's telling this story. He's eyeing his fourth helping of venison, wiping the corner of his mouth with his napkin, and he starts to giggle. "The wimpy sonofabitch says, 'What's the bad news?' So I have to tell him his cabin burned down. There's a long silence, so long I think he's gone away — he's not taking it well at all — and finally, after the longest time, he's able to ask, 'Well, what's the good news?' and I get to tell him" — Mike's laughing, wiping his eyes now, delighted with himself, free at last — "I get to tell him, 'Well, your pipes thawed out.'"

We're all laughing, all of us lucky, jealous caretakers at the table. Damn landlords!

"Pansy sonofabitch," Mike says, still grinning, his mouth full now, chewing. "Couldn't take a joke. I never did like him."

There are ghost stories, too, about this valley, and the one I know involves our ranch. Fix Ranch is named for Edison Fix, who homesteaded the place (Fix Creek, which flows through the ranch, is also named for him). He lived and died here on this land, and loved it, and never left it. Mr. Fix requested that he be cremated (he lingered with cancer here in the greenhouse, in this shed where I write, which used to be his cabin, back in the thirties) and that his ashes be sprinkled over the ranch — not from an airplane, but on the ground. I'm told his widow just cast his remains around like fertilizer, like grain to chickens, but perhaps, who knows, she put them all in one special spot.

We've never seen old Edison Fix, but back when Holger

used to run guide trips out of the lodge, hunters staying in the guest cabins would see him now and again. We've even had some friends stay over who've mentioned being extremely frightened deep in the night, upstairs in the guest cabin — country friends, too, not city friends. The next morning they'll say they've heard loud, angry noises, like someone slamming something around, and footsteps coming up the stairs.

The first couple of weeks I was here alone, before Elizabeth took the train out, almost every night I had the nightmare that someone was coming slowly up the stairs, someone angry — and me, in that terrible dream way, unable to move, unable to make any sound, any protest. That person, that force, was an old man, and he would sit on the side of the bed. On one bad night — and I felt this surer than I've felt anything — the hand of this person sitting on the bed grasped my ankle, and though he did not twist it, he would not let go, either. I assure you I am not talking about being tangled up in the covers, because it was early September, and I was sleeping with only one light sheet over me. Besides, there was more, more than even the hand's tight grip: I felt a hair-raising, blood-backing chill of evil in the room — the electricity of it hanging like a loud echo, but also growing, increasing, like a dog's heavy breath, getting worse and worse, more and more ominous.

These things happened. I'd passed them off as nerves or stress or the strangeness of a new land, until Mike and others told Elizabeth and me about the hunters who kept saying they saw Fix's ghost — the ghost of him as an old man, which must have been the time when he loved the land, the ranch, the most.

I've never seen his face and don't believe I ever will. I think our peace has been made. I think the ghost — the force, the energy, the leftover feelings old Fix had for the valley — is appeased. He, or it, knows that we'll be staying, unlike the hunters who come for a week to shoot an animal and then leave, fleeing the coming winter. There have been times, though, when I've gone for a walk at night up in the woods behind the house, on an old logging road covered with a canopy of tall, lacy-leaved cedars and larches — up past the pond, too far into the woods — when I've felt something, someone, behind me. I'd turn around to look back down the trail — the old logging road shot blue with bright moonlit patches and hard, dark shadows — and it'd be clear that there was no one there. Yet I could hear him, feel him, sense him, standing in the middle of the road, watching me, looking right back at me, like an animal — hands on his hips, perhaps, and a strange badness in the air. Maybe he was frowning, as jealous of my being alive here as our Yaak Valley landlords, scattered across the country, must be of us caretakers, who are living this life instead of just owning a piece of paper, a paper title to a property, as in a Monopoly game . . .

We're here, we're alive. Fix isn't anymore. Of course he's angry.

There are days when I promise, when I swear, that as long as I can walk up the trail behind the house, or as long as I can go out into the yard and look up at the stars, I'll never be unhappy, never. Not just count my blessings, but shout them.

That night of the dinner party, the one where Mike told landlord stories, the northern lights — green and yellow — came out around midnight, after everyone went home. They shone over the tops of the mountains like the announcement of a strange Broadway opening. Then they began to shoot, in streaks and flashes — and rolling, too, like fast rivers — across the sky, right over our heads. Elizabeth and I stayed out and watched them for an hour and a half, before they faded and then disappeared.

This valley shakes with mystery, with beauty, with secrets — and yet it gives up no answers. I sometimes believe that this valley — so high up in the mountains, and in such heavy woods — is like a step up to heaven, the last place you go before the real thing.

October 5

My chain saw is a Stihl 034. It's orange and white, made in Germany, and it's not a professional's saw, but almost: it's big and strong, a wild horse, and I'm scared to death of it. I've read and reread the sixty-page instruction book several times. My Stihl starts like a fine car, fuel-injected, and rumbles nicely with a throbbing, rattling purr when it's set on slow idle. It's complex and has all sorts of tricks and gimmicks, such as carburetor preheating, which is required at zero degrees and below (most people cut wood in the winter, so that it "warms twice"; you work up a sweat sawing and then get warm again burning it). The manual uses phrases like "blipping the throttle," and is chock-full of sentences like: "A Rollomatic guide bar is not necessary to separately lubricate the sprocket nose bearing because the chain oil which flows to the bearing by way of the bar groove during normal operation is adequate for lubrication," and "This is necessary because high contraction stresses would otherwise occur as the chain cools down to ambient temperature, especially at extremely low outside temperatures, and cause damage to the crankshaft and bearings."

I'm a slow reader, even with things I understand. I trace the words over and over again with my finger, saying some of the sentences out loud. Elizabeth believes that this concentration on and fascination with the owner's manual is boy stuff, a guy thing, that I'm enjoying the schematics and such — the hard-edged rules and facts — but it's not that at all. I'd rather be walking through the woods or sitting by the river reading poetry, but this is something I've got to learn. This is something that is going to keep us warm.

There's something in the manual about "whitefinger." I've heard people in the saloon talking about this condition, have seen guys who have it chronically. It's amazing they're still able to work. Their fingers are swollen and white, scaly and flaky, and they can hardly move them; they stare at their hands while sitting in the bar, holding a drink, or reaching for a mug, as if the hands do not belong to them.

"Prolonged use of chain saws (or other machines) exposing the operator to vibrations may produce whitefinger disease (Raynaud's phenomenon). This phenomenon reduces the hand's ability to feel and regulate temperature, produces numbness and burning sensations and may cause nerve and circulation damage and tissue necrosis."

The way to avoid whitefinger is not mentioned. I suppose I should rest and exercise my fingers every so often, get some blood flowing into them; I should cut a cord or so and then set the saw down, to rest it and myself — I need my fingers to write with — and go for a walk up into the woods. If there were not a disease like whitefinger, a danger of it, I would have to create one.

I've got a hard hat, goggles, and earplugs; steel-toed Red Wing boots and leather gloves. I'm scared to death of this big saw, and yet I'm excited about using it.

OCTOBER 6

I went to cut wood today. The air was stiller than I've ever felt it. I sat in the car and read the newspaper I'd bought at the mercantile. I read it slowly, with the sun coming in through the dusty windshield, while leaning back in the seat, sleepy, reading it like a rich man with nothing to do. Then I put my dust mask on, my helmet, gloves, and earplugs, and cut wood for over an hour: big logs, with the newly sharpened saw, so that the chips were flying as if in a blizzard or a ticker-tape parade. It was a mild day — such a blue sky! — but I cut and cut, until I was dripping with sweat.

Afterward I climbed down off the top of the slash pile — stepping gingerly from lodgepole to lodgepole, picking my way down the jumble of toothpicks — and took off my helmet, my shirt (the air cool and wonderful on my dirty body), the dust mask, and last, savoring it, the earplugs.

It was silent, even without the earplugs: no wind, no noises.

Later, as I was splitting the wood, and loading it into the truck, I heard one raven. I won't say it had been frightening before I heard that raven, but it had been very lonely, in both a good and a sad way.

I'm so lucky that Elizabeth likes it up here. What if she didn't? What if she were unhappy? What if she wanted to live in a city?

After a while, the raven stopped calling, and I was alone again.

OCTOBER 8

Almost midnight. The woodshed is floating in magical wood. I'm hoping for a cold spell, the hardest winter on record.

I've been taking my shotgun with me when I go up the mountain for wood, in case I see a grouse. Yesterday I saw one. I got out of the car and followed it into the woods, running, and flushed it, shot it as it was flying, golden, into the sun — a heavy ruffed grouse, like a villain in one of Shakespeare's plays.

Elizabeth and the dogs were beside themselves when I came home with both a bulging carload of wood and a grouse. We cleaned it at dusk, shivering, saved the prettiest feathers, and fed the feet to the dogs, who were waiting to be included in the kill. We bought wine and will have the grouse tomorrow night.

Cloudy tonight, with a wind out of the north. My back's sore from cutting. Clouds blowing past the moon, no stars visible, dark clouds blowing past. I took the dogs outside and when I came back into the cabin Elizabeth asked how cold I thought it was.

"Cold," I said. It was about ten-thirty.

We sat there and thought about how much colder it was yet to become. It was a strange and scary feeling knowing that it, the great cold, lay out there in the future, in the dark, a certainty.

The Falcon, the good old V-8, pulls tons of wood, gliding up and down the hills with it, through the mountains, along ridgetops, looking down at the river far below. O Falcon!

A woman in Yaak was calling her husband Numbnuts today. I was on the phone to Texas, clearing up some things about a prospect my geology partner and I have going, an oil well in Alabama, and the woman was right next to the pay phone when she bellowed across the street to her husband, who was standing outside the Dirty Shame, shooting the breeze with Truman: "Hey, Numbnuts! Come over here!"

"Wild country," I explained to my partner, an older gentleman. "A hard country."

And I pictured him repeating this later on, to people in the Petroleum Club, over lunch: how there's a woman in Yaak Valley who stands in the middle of Main Street and shouts "Hey, Numbnuts!" when she wants to get her husband's attention.

October 9

We're gleeful it's so cold! I don't know what the temperature is — we can't pick up any radio stations — but it feels the coldest yet. I'm guessing about fifteen degrees. I had to hold the pen over the stove this morning, to warm it so that the ink would start flowing: just a regular old cheap ballpoint. (Larry Levis, a great poet, uses a black Parker; I'd forgotten about that until just now. Anything Larry Levis says to do, about writing, I'd better do. And now I'm remembering Jim Harrison's advice: black pen at night, gold pen during the day. Anything Harrison says, likewise.)

Last night it was cold in the house. We fell asleep on the floor in front of the fireplace, warm after our lemon grouse (dripping fat and juicy), dirty rice, homemade bread, baked potatoes, and the season's last salad. This morning it was not warm. I fixed coffee on the propane stove in the kitchen, and held my hands above the hissing, spouting, tiny blue flame as the water heated. I didn't want to take them away when the water started to boil. I wanted to stand there all day.

The pond across the road was frozen for the first time.

O Falcon! While the Japanese are creeping about, punctuating their spells of hibernation with occasional flurries of lethargy, searching for a replacement transmission — for eleven days now, they haven't been able to find one; not in Denver, not in Seattle, not in the United States, not in Japan; the Nissan has been in a coma — the old Detroit V-8 has been pulling heavier, greater loads up and down the mountains.

The springs are shot, the frame is scraping the pavement. I drive home at fifteen, twenty miles an hour with wood lashed to the hood, to the roof, and the trunk sprung open as if a car bomb has gone off. I've got the windows down, extra logs are hanging out the windows — wood in the front seat, wood on the dash, wood in the glove box, wood in my lap — the car groaning back and forth to Hensley Mountain, the little Falcon, never before called on to be even a passenger car, not in the last ten years, anyway. Valiantly she fills in, struggles on, knowing the severity of my predicament.

I want to go over to Glacier before the park closes for winter, but to take her on a vacation might be asking too much.

The old Falcon gets dirtier by the day, holes poked in her upholstery, chain-saw oil stains on the seats, wood bark ankle-deep on her once-immaculate floorboards . . . but there is no other option. It's do or die, and she's doing.

Scab wood-getting. The Nissan is on strike while in the meantime winter advances.

Midnight. Two cords cut. I carry the wood up from the bottom of the hill, load it into the Falcon, drive back home (four trips), unload, split, and stack it, then go back to Hensley for more. There's nothing that gets you as dirty as woodcutting. I thought roughnecking was bad. Roughnecking is a tea party, ginger cakes and lemonade. Roughnecking is washing your hands and blowing your nose with a lace hankie, compared to cutting wood.

I go through a pair of leather gloves every day. The steel teeth of the saw's chain need filing and sharpening every night.

My clothes cannot enter the house. They are too soaked with oil, sweat, and gasoline, and caked with sawdust. It's in my beard, my hair, down under my shirt. The dust from this, the driest year on record, is everywhere.

Like a swami, like something immortal, from another planet perhaps — Mississippi — the Falcon glides through it all, sometimes obscured by the clouds of dust it raises. Like the *Peanuts* cartoon character Pigpen, the Falcon and I are still getting wood.

My knee is blown out and stiff from where I whacked it with a backswing of the maul today; bruised ligaments only, I hope, rather than torn. I went down with a scream and stars in my eyes, twisting halfway around and falling on my back. I came to a few seconds later to see clouds, lying on my back and looking up at the sky like a dead actor.

There are blisters on my feet from trudging up and down the hills in my steel-toed boots. I have splinters where wood has stabbed through the worn-out places in my gloves. But I'm going to hang in there with the Falcon if I can. I'll bet we moved twenty tons of wood today.

Fingernails split. Whitefinger. And I have escaped the real injuries.

OCTOBER 10

I know *nothing* about this valley. Everything I see is new, and I understand nothing. In the daytime I'm not frightened — there's too much to learn — but at night, when I fall asleep . . .

This valley, this landscape, has the option of taking me or

rejecting me. I don't have any say in the decision. It's as if something preordained has already been decided — either I will fit the land or I won't.

It has everything to do with people, as well as the land, up here — even though there's only thirty or so of us. It's like we're a *herd*.

There are forces in the woods, forces in the world, that lay claim to you, that lay a hand on your shoulder so gently that you do not even feel it: not at first. All of the smallest elements — the direction of a breeze one day, a single sentence that a friend might speak to you, a raven flying across the meadow and circling back again — lay claim to you, eventually, with a cumulative power.

Tim Linehan had tacked a note to our door — the pencil marks smudging on a torn-up fragment of grocery bag — inviting us down to his and his girlfriend Joann's lodge on the shore of Vinal Lake, which they're caretaking. When we got back from town, the note was flapping in the wind.

The larch trees are turning. They're the first to turn.

We heard a coyote yipping in the meadow yesterday afternoon, barking in the wind. The tops of all the trees were swaying gently, but when I looked at the wall of the forest, it seemed that the trees were all standing at attention, standing there watching me, and barely moving — just up high. Cottonwood puffballs blew through the field. The woods will either have me or they'll send me home. Every small sight, every small action, counts. That coyote's barks are accumulating, becoming part of my life, and I am turning away from my old life and walking into a new one; Elizabeth and I both are.

Of course we're homesick. If it came cheap — our happiness and freedom — it wouldn't be worth having.

Tim said to bring bathing suits if we wanted to swim in

the lake. But isn't it too late in the year? Isn't the water too cold?

The days are fine. It's the scary dreams, dark dreams, at night that get me. In the day, walking around breathing the clean thin air, seeing nobody, only Elizabeth, and feeling the days change — it's like a daydream, a lazy, lovely feeling, a great dream, one I seem less and less able (or willing) to awaken from.

Only at night do the doubts come.

I need a lot more wood. I talked Elizabeth into coming up here — this far. The least I can do is get enough wood. Or try to.

OCTOBER 11

Purple thunderclouds over the valley, and new friends, Tim and Joann. We had grouse breasts, gotten from the woods above their cabin, huckleberry cream-cheese pie, and white wine. The wind blew whitecaps across the lake as we sat at the picnic table and watched the gold larch needles fly through the air. You tell me, on a good day, is there something better than this life?

Joann is from New Hampshire, like Tim, and owns a house there. "My parents keep asking me when I'm going to come home," she said.

OCTOBER 14

"Snow clouds," I said to Elizabeth. We'll see.

Patches of sun shine on the hay, on the late-season dry grasses; green logs pop and bang against the sides of the old tin stove in the greenhouse, where I'm writing. I'm thinking about how I love getting wood. I moan about it, and oversleep some mornings when I've worked too hard, gone beyond my ability, my capacity, burned myself down at it — and yet I really, really enjoy that feeling: lifting the great logs, loading them into the car, bringing them home, splitting them, stacking them, the pile growing higher, the fortress, the protection against the cold, the currency of the North already down in the woods, waiting for me to pick it up, and free for the taking. I'm sure a psychiatrist somewhere will be reading this and clucking his or her tongue, going "Tsk, tsk, tsk," but that's the way it is, the way I feel.

October 16

Today was a warm Indian summer day, but there'd been a frost in the morning, and there was a breeze in the tops of the trees to remind me of what's coming. I went out, as ever, to cut wood. I wore my oldest clothes, as I always do — no need to ruin new ones on such a brutal, dirty job: my old boots, my faded overalls, and my soft old brown long-sleeve shirt.

I drove up a logging road with the CB on, to listen for runaway hell-bent logging trucks. I didn't hear anyone, and when I came to my slash pile, I got out and sawed for over an hour.

After I shut the saw off and began loading the logs into the truck — the woods ringing again with silence — I heard, and then saw, two pickup trucks coming down the mountain, driving fast.

They passed me — scraggly-looking customers in both trucks who'd perhaps been out scouting for the coming hunting season (they didn't have any wood in the back of their trucks).

I'd forgotten to turn my CB off, and they didn't know I had it on. I stood there sweating in the mild sun with my helmet off (my hair wild and thinning, unkempt), bits of wood bark all over my shirt and overalls.

They talked to each other on their own CBs as they rattled past, after they rounded the bend.

"He looks kind of funky, doesn't he?" one of them said. My radio crackled, reporting this news to me.

There was a brief pause, and then the other answered.

"Yeah, he looks *funky*," he agreed.

They didn't say anything else, and after a while I was alone in the woods again. I pondered, for the rest of the day, what "funky" meant.

Later in the evening, before dark, I hiked across a small clearing on the mountain behind the house. I came across two piles of grizzly scat, big mounds of it, the scat as thick around as my wrist. It was fresh. The bear had gorged on berries.

It was dusk, so I turned around and headed home, humming, to let the bears know I was coming through their woods. The woods took on their full meaning. I didn't hurry, but I didn't dawdle, either.

OCTOBER 18

The days are falling back, going so quickly, and yet each one is filling us up so full. Last year I didn't know what a merganser was, and this year there is one on the pond across the road.

Everything's going on, back in the woods behind the house. I found mountain lion tracks in a puddle. When I look at the trees, they're still standing the same way, waiting for winter. *They're* ready.

Late in the afternoon, there's a period when the light turns so strange, so bronze and still, that it's like a tintype — as if it's trying to hold that angle of light for as long as it can, for us to look at the fields and woods and meadows in that sharp light one last time before falling away. One last time . . .

And we look. We stand there, in honor of the light, and just look. Birds call back in the woods, flickers and thrushes, and my life seems about to *speak* to me, that sense of waiting, of promise, is so strong.

OCTOBER 25

The sun is shining and the rain is falling. We've all seen this happen, but what I mean is both: it is a bright, perfect day, and a fine but heavy rain is falling from a single cloud that's floating meekly across the valley, in no way impeding the bright sunlight all around. A warm spell. The dogs' water did not even freeze in the red metal bowl. I think I am beginning to see a pattern, discern through this rain and sun — that there is no pattern.

The elk hunters came to the valley today, including the landlord and his fiancée. Holger was tired from traveling (he'd been in Miami Beach), but not so tired that he did not have a quick survey of things that need doing immediately. He wants the wood chips up off the ground, where I have been splitting logs for the stove and fireplace. He wants another caulking on all the logs around all the buildings; he showed me where a fifteen-gallon drum of tar is, back in one of the garages. He wants the batteries taken down to the mercantile to get them charged a little stronger.

I'm inside the greenhouse — always inside the greenhouse — writing, and I resent his presence, even though it is his ranch.

The fields are turning their yellow hay color.

I know there will be a new list of things to do, some of

them expensive, when the landlord gets in this evening
from his first day of hunting.

He is a great guy, and so is his fiancée. It's just that, well,
can't he see I'm trying to write? So what if the place slips,
goes downhill a little?

All around me, the wind is swirling the golden larch
needles past the greenhouse windows.

OCTOBER 27

I'm falling away from the human race. I don't mean to
sound churlish — but I'm liking it. It frightens me a little
to realize how much I do like it. It's as if you'd looked down
at your hand and seen the beginnings of fur. It's not as bad
as you might think.

I have a family back in Texas — my whole family, for five
generations now. Elizabeth's whole family is in Mississippi.

I need a dream hoop. The days are great, heavenly. It's
only at night, sometimes, that I feel a little guilty, and the
nights are long at this time of year. No one in our families
has ever left home, not for good.

They're so far away. My family, and the other part of my
life — friends, acquaintances — they're another galaxy. It
makes no sense, it's a strange way to live, but I like it.

If happiness were cheap, it wouldn't be worth having, I
tell myself — again.

OCTOBER 28

Holger got an elk. I could tell he had by the way the dogs
went berserk and began doing flips and cartwheels. I went
over to check the back of his truck, and sure enough, it was
all there. Tomorrow will be butchering day: bones, bones,

bones. The dogs could barely get to sleep, they were so excited — like wild puppies.

The generator was popping and punching around very
badly when I got in from wood cutting and started it. I
hope it's just the plugs. They looked clean when I checked
them, but they almost always look clean on a propane-
burning engine, even when they're worn.

Holger got his elk, but I didn't get my truck. Unbelievable. The Nissan people wouldn't give me anything on my
trade-in — the truck they'd had for forty days and nights
and had still not even begun to fix. I mean, they offered
zero in trade, so I insulted their central office, stormed off,
and went straight to the Toyota store. I spent three hours
negotiating there after I'd pulled my sorry Nissan over with
my rusty old tow chain, for the Toyota people to inspect
and evaluate. Picture it: a car with 150,000 miles pulling a
truck with 180,000; two hound dogs; a shot and completely
disassembled transmission stacked in a box in the back of
the truck; Mississippi license plates on the car, Texas plates
on the beaten orange truck. After negotiating to the bitter
end for a new Toyota, and not being able to make a deal, I
wound up telling the salesman that he had dirty, crooked
teeth. I left for Glacier National Park, where Elizabeth and
I walked around Lake McDonald with the dogs, skipping
stones and brooding about money.

Tomorrow a respite, getting wood and typing and perhaps helping Holger butcher his elk, and then Friday, a
journey into Spokane. I know the truck I want. I know
what color (red or black — what I have learned the salesmen, these days, call "hot" colors. Spare me).

Stopped off at a clearcut on the way in, and with the
headlights shining out across it, the engine idling, I walked
through the cut-over, beneath the hard stars and mackerel
clouds and quarter-moon — a warm night, not even freez-

ing — and picked up a few armloads of loose wood for tonight's fire, and tomorrow's breakfast, and tomorrow night.

It's cold upstairs. We're warming slowly beneath the blankets. This big house gets cold. I must get tougher. Also, I must remind myself that I don't need that red truck.

I'm still a little shocked, angered, and bitter, but mostly incredulous. I was offering them a lot of money. They wouldn't take it, though.

Sometimes I get mad too easily. But sometimes it feels good: holding back and holding back and then finally just saying oh what the hell, and letting them have it.

Salesmen. The bastards.

Snow's coming on. I've got to have something reliable.

OCTOBER 29

Walking the long walk down the empty road to visit with Breitenstein, to learn some more things. Walking past the sweet-smelling mountain ash, yellow leaves on the rough gravel road, and walking past Fuel and Buck, who raise their heads and watch me. Later, crossing the narrow bridge over the south fork of the Yaak River — late afternoon, autumn blue sky, alert for grizzlies and moose in the willows along the river — I finally get to Breitenstein's house, but he's not in.

I sit down to rest in front of his big barn and look all around at his ranch. It's an oasis of control, of order, in the wilderness. The chinking between the logs of his outbuildings is whitewashed. His large lawn is mowed. There are no stumps in his small meadow, which has the greenest grass in the valley — though he owns no cattle, only a meadow; cows are too messy!

The barbed-wire fences surrounding his meadow are

razor-tight, straight-edge perfect. His old dog, who trots up to greet me, does not approach past a certain point — unwilling, doubtless, to cross some imaginary line that Breitenstein has confined him to — and it's only the dog that's *not* perfect: one eye is cloudy, milky blind, though the other eye seems fine.

I sit there with my back against Breitenstein's barn door for a while, waiting, but there's too much daylight left. He must still be out in the woods somewhere, trimming branches perhaps, tidying up, and so I sit a little longer, and continue to look around at his sharp ranch, at his perfect fences, and I try to pretend that I am he — which is what a writer does — and then, when it doesn't work, I get up and walk home.

His dog follows me to that imaginary line, but then stops and watches me leave.

OCTOBER 30

Damn, I need a haircut. While test driving another Nissan today in Bonners Ferry, Idaho, I told the salesman I was a writer, and he looked at me and asked, "What, science fiction?" I do look like a nut, a lunatic.

OCTOBER 31

It's a dark, slow morning. Purple clouds over Libby, over the Kootenai. It's not cold at all, but I still wore a light sweatshirt, walking out here to the greenhouse; I've got a fire going too. The window panes seem funny this morning: clearer and somehow thinner than usual. Everything looks still, the kind of stillness that seems perfect when you just stare out at it, hearing nothing except for the ocean noises in your heart. But then you focus on something —

an individual stem of late-season wheat, perhaps — and you see that all is not entirely still, that there is a slight breeze along the ground, that the smaller grasses wave slightly, but only if you focus on them. Everything else is still.

Rain down in Libby, forty miles away. Will it turn to snow? The salesman in Spokane was a turd. Fortunately, we didn't even have to shake hands.

Halloween in Yaak is this: strap on a pair of deer or elk antlers, go down to the Dirty Shame, and drink a six-pack. No glitter, no makeup.

Ranier. Schmidt. Olympia. Bud. Michelob.

In 1983 it snowed on December 8.

In 1984 it snowed on October 27.

In 1985 it snowed on November 7.

In 1986 it snowed on October 31 (a foot and a half).

I would like to win the snow pool. I would like to buy a round for everyone at the saloon, and then buy firewood with the remaining money.

I did not get all the wood in. I dallied in Libby, some days, when I'd go in to make photocopies, and I jacked around and didn't buy a new truck soon enough. I've got a deer hunt coming up, with my family in Texas. I've got some wood in, a lot of wood — eighteen, maybe twenty cords — but I'm burning some of it every day now out here in the greenhouse and in the kitchen stove. I didn't get the thirty cords I wanted to have by this date.

Writers. Half-assed at everything, it seems, except, occasionally, their writing.

I'll go get another load today. Tricky, of course, because (as ever, the days rushing by so fast) it's a weekend, and I don't want to start the saw up in the woods, don't want to ruin the deer and elk hunting for others.

I know how I'd feel if I were up in the mountains, watching for deer, and heard a chain saw start. I'd wonder why the guy didn't get his wood in by August or September, as the larches were turning gold and the leaves were blown by the wind, tumbling across the roads.

Yellow cottonwood leaves rise knee-keep on both sides of the road leading to Hensley. A lot of the trees are getting bare now. This is like a marriage. Something is coming on. I hope I can do this.

NOVEMBER 3

Football at the Dirty Shame last night, Dallas versus the New York Giants. A good, fierce game, but we left at half-time. How old I've become! No job, I can stay up all night, sleep 'til noon, go for hikes, all of it — and I wanted to get back and do some typing.

The greenhouse smells funny, only I'm not laughing; some large mammal was in here not long ago and probably left at daybreak. It smells like a wet dog — coyote? wolf? — and it's pretty strong, not the smell I'm used to. It came in through the roof, I think.

Last night the men were distressed at the warm temperatures. Jimmy, one of the bartenders, said he got up at two in the morning and the temperature was only forty degrees. Artie, another of the bartenders, said he'd checked his thermometer at four A.M. and it was only thirty-four degrees, and he's five hundred feet above the valley. Insomniacs of Yaak.

Mostly just the bartenders watched the game, and us. The hunters were too tired and had headed home. Suzie looked a little wired, a little edgy, and awful glad to see us. She was real friendly.

I'm going into Kalispell one more time to pick up the old

orange truck — they finally fixed it — and to try, one and only one more time, to buy a new one.

I had nightmares all of last night, in which various salesmen insulted me, goaded me, and then I got into fights with them, actually assaulted them, and ended up losing in court and going to jail.

A windy, warm morning today, the grasses lying almost all the way back flat. There is nothing as exciting as the wind. New love — and then the wind. But the wind has always been there. Even before you knew of love, you knew of the wind. The wind could excite you as a child, and it still can, and will.

Slipping my last sweatshirt off now, writing shirtless in November. A strange country. The landlord assures me I will not be able to write straight through the winter here in the greenhouse.

It's not a particular goal of mine to do so. If it happens, it happens. I will have plenty of wood just in case. I will try to.

The wind is blowing hard from the north. A fine mist is sleeting, and I step outside the greenhouse to watch it, to feel it. When I go back inside, I am surprised at the woodshed's smell, like a smokehouse, and its warmth. The sleet is bouncing off the old shingles, spitting against the glass.

I'm surprisingly calm. I don't know if this is it. But I've got almost all of my wood in. I'm about ready to get on with it, if it really is winter this time.

What a good sound the wind is making. What if this is the sound you hear after you lie down and die? I am not ready for that, but I am ready for this.

NOVEMBER 4

We drove to Kalispell, the big far-off city — the nearest McDonald's is there, three hours from our cabin — to do truck shopping. We stayed overnight at the Blue & White Motel ($20.95, winter rates). I've made what looks like a deal for a new truck, and will try to make it final (not "finalize," which to my mind is *not* a verb, is not even a word!).

The thing Elizabeth and I agreed on before going to bed was that we haven't missed shopping malls at all, living up in the Yaak. They're even tinnier and more ludicrous, after an absence from them, than what we had initially thought back in the city — and our opinion then was considerably low. The air in them is bad, the lighting false, and it feels as if the people running the cash registers in those little booths are taking something from you besides your money, when you finally get to ante up.

And this morning in the motel room, curious about the weather, we turned the television on briefly, with the volume down, and looked at Willard Scott's map. I didn't have my contacts in yet, so I thought that over Maryland ("Delightfully warm" on the map) it said "Godalmighty warm," and where I thought it said "fog" was really a temperature, 109 degrees. But even with the sound turned down I found the shape of Willard's mouth weird, his eye movements crazy, and his facial expressions insulting — too much to bear, even for a few seconds. So I shadowed him with my hand, screening him from our view with my palm, like a basketball player setting a pick. Every time Willard moved, my hand moved with him, and we got through the weather okay, though we learned nothing that we didn't know already, that it was warm in the South and cold in the North.

The belt of cold air that came through the greenhouse

the other day showed up clearly on the map, a downward nose, the western edge of which passed right over the flanks of Yaak. That cold air hit the warm air yesterday morning at ten o'clock and turned the warm air back around and sent it south. It took Willard twenty hours to find out about this, and show it to me on his map. I felt smug and independent.

November 5

Camping alone. Bundled up tight. Drove up an old logging road ten miles, climbing, twisting; crossed over the pass and into Idaho just in time to see the sun setting behind the endless chain of the Kootenai Mountains, more saw-toothed and spectacular even than the Sawtooths. And quiet. No one up here.

British Columbia lay five miles up the road. All the valleys below me were filled with a pale, luminous blue fog — glowing. The mountains themselves were a shining cobalt, and the sky was a wild and lurid red and gold, a sky evangelists must see each night when they go to sleep and dream.

I ran a bull moose across the state line, out of Montana and into Idaho, where he'll be safe, where there isn't a moose season. He was walking down the road in front of me when I went around a curve, and he leaned forward and broke into a trot, staying ahead of me, and then, as if getting annoyed that I was still following him (I had to; my mouth was agape; I had never seen an animal that large before), he broke into the smoothest of gallops, a lazy, long-legged floating. His wide cradle of antlers could have held a tea service without spilling a drop, so smooth and level was his gait.

The road had a high cliff on one side, and alders and a

steep drop-off on the other, and he couldn't go left or right. So he ran down the road in front of me in his easy lope, with me not chasing him but following at a distance, and the great flap of muscle on his neck shook like a camel's hump, flopping back and forth. I could tell by the stiffness and hesitancy of each step that he wanted to turn and be chasing me but was not sure how to go about it, how to reverse the process.

We were about a quarter of a mile from the Idaho line (a post, unlabeled and unnumbered, just an old wooden post, marking the spot), and shortly after we crossed over, he spotted an opening in the alders on the down side of the cliff, ducked his head and antlers into the opening, and like a good fullback disappearing into a small hole at the line of scrimmage, he delicately and quickly muscled his way into it and was gone.

I listened to the crashing of small trees, of alders being split. I had some sandwiches in the glove box that Elizabeth had fixed for my supper and for my breakfast and lunch tomorrow. I was so excited that I stopped right there, turned the truck off, and ate all those sandwiches. Then I ate a whole package of Rolaids. I had my rifle in the truck, and I told myself that if I saw a deer in Idaho, I was going to shoot it — I had an Idaho license — but shooting a moose would be like shooting a cow, or a billboard.

Camping, then. Full moon. Frost already, at seven-thirty. I heard the distant sound of a train, the Union Pacific, going up through Moxie Springs and into British Columbia. I imagined hoboes were riding it, also in their sleeping bags, like me. Elizabeth was glad that I could get away camping and hunting, and I know she's glad to have the house to herself. A fire going, and a book. The dogs asleep next to her. We are going to make it up here. Whatever it takes, we will make it work.

N O V E M B E R 16

It started snowing today, in the early afternoon — we'd heard it forecast on our return trip from town, and were hurrying to get home before it came down. The sky has been full of snow clouds all day, and there was no question that it would snow. We only had to wait.

We fixed hot chocolate, sat by the window, and read, and waited. We went for a walk in the woods with the dogs. We didn't see any deer or squirrels or even any birds. I've got a lot of wood in, but could use more.

It was like guests showing up, when the snow first started coming down: guests from out of town.

I'm from the South, will always be from the South. I'll never get used to snow — how slowly it comes down, how the world seems to slow down, how time slows, how age and sin and everything is buried. I don't mind the cold. The beauty is worth it.

It's dark now, and still snowing. An inch, two inches, on the ground, on the car, on the trees, everything. If anything needed doing before the snow came, it should have been done yesterday.

It's going to be strange falling asleep tonight, knowing that snow is landing on the roof. It's here. We're here. Nobody's leaving.

NOVEMBER 18

I hiked to the top of Flatiron Mountain today. I got up near the tree line at dusk, thrashing around in butt-deep snow, and flushed a family of fat grouse from their lodges in snow-crusted Christmas trees. I kept going up higher, into the fog clouds, sweating like a race horse from the steepness of the climb. It's a barren mountain, because each year avalanches wipe it clean like a slate. It makes me dizzy to know that Indians used to cross this special mountain, and after them mountain men, trappers — anyone in the valley who's ever been wild, who's ever been worth a damn, has hiked up to the top of Flatiron at one time or another in his or her life. Probably not many have gone to the top in deep snow, and that thought somehow made it seem cleaner, newer.

The sweat froze on me when I reached the top; it took me two hours to climb. The patchwork maze of hideous square and rectangular clearcuts in the mountains across the valley behind me disappeared once I got into the wind and fog and mist, into the clouds — just me and the Indian spirits, me and the mountain men spirits. I had no reason for being up there at dusk, no reason for being up there at all, and yet I had every reason.

It took me only one hour to get down, running, half skiing, sliding on my back like a turtle, down and over the steep edges and cliff faces, protected in my wild falls by the deep snow, its softness, and no one around to see me, no one in the whole world. The joy of running across and down the vast snow fields in that fog was like running down Olympus — Zeus as a boy of seven, perhaps.

I had told Elizabeth where I'd be, and what I was going to do, so that if I broke a leg, she could send a search party — dogs, helicopters — up after me.

This is the only mountain safe to cavort on in hunting season. It's too high, too deep in snow, and too steep for hunters or game — even if the hunters did see game, it would be too far, through the snow, and too steep, to pack it down. Sixty- and seventy-degree slopes all the way up, all the way down. It's safe from the logging companies too, because there aren't any trees; it's all rock that's been scored clean by the avalanches.

I was dizzy with excitement, dizzy with the beauty. It's not avalanche season, but I could picture what one would be like: the whole mountain beginning to slide beneath me, coming to life.

I am coming to life myself up here in this valley; I've

taken a broadax to my life. Zeus, at the age of seven. I know I won't live forever. Why does it feel as if I will?

NOVEMBER 19

We're unable to get used to the snow — the way it comes and goes, for example. I was standing around the mercantile when it began snowing again, and noticed that I wasn't the only one affected this way. People who've lived up here for ten years or more also stop what they're doing when the snow starts to fall, and say, "It's snowing again."

There's about a foot on the ground. It snowed hard the last two nights. The snowshoe hares are white and safe. The hunters are having an easy time tracking the deer and elk. No big elk have been taken, but there've been some deer killed, big bucks, over the last couple of days. You can sit on the bench outside the mercantile (in your heavy coat, sipping coffee) and wait for hunters to come driving by. It's sad to think that the big deer they're bringing in lived to see another season's snow, but just barely.

But there is no capacity for sadness, no way it can linger. There is only room for excitement and wonder. Perhaps this is what, finally, gets the big deer killed. It's strange. I'm a hunter, but I can't help imagining that the deer feel the same way we do, when that snow first starts landing on their backs, on their antlers, and everywhere.

NOVEMBER 20

Already I'm letting myself go a little. I try to shave every other day, but lately I let three or four days go by, sometimes four or five. I sleep on my hair all wrong too, and it's drying out up here, getting all wiry and crazy. Elizabeth doesn't seem to care, and she never mentions it. She misses electricity, and National Public Radio — *any* radio — and even television (it's good for couples' tastes to vary a little), but still, we seem to spend all our time celebrating the things we love rather than lamenting, for example, that we don't have a washer and dryer.

Town — Libby — is for laundry. There's a hunting lodge in Yaak Valley that's got a washer and dryer for its clients, but I've heard that the well water contains minerals that turn all your clothes red.

No radio stations are able to make it in over the mountains and drop their signals into our narrow valley. Friends offer to send us special shows they've taped — bluegrass, jazz, and rock — but they don't understand the loneliness these things would create, and they don't understand when we say no, but thanks.

That's the greatness of live radio — not the mind-filling soulless chatter that's used mostly to plug up the void, but the old-fashioned kind, the kind you still get on late-night drives across the country; you can turn a knob, twist a dial, and make contact — *contact* — with another living human being.

But to have the artificial, stale echo of such a thing, the *ghost* of that living human being — sitting up in our cabin at night, trying to pretend that that voice, those music selections, are spontaneous, living in the same moment as us — well, that's lonelier than silence, it underscores our isolation, and so we say no thanks, and listen to a few of the same old tapes and albums over and over again — rock, country, classical, anything. At night when the generator is running, our cabin is filled with music, but we do not try to imitate radio.

And television! Well, Elizabeth does not miss it that

much. She sews in the evenings, we play Scrabble, we play pinochle (our own special brand, where it seems we'll never stop; I like the game okay, but she loves it; she's already fitting into the valley, in some ways better than I ever will). Or we'll just lie by the fire with the dogs and pet them. It gets dark so early.

Neither of us misses a telephone.

Listen, I've found out, to my great delight, that you don't need one. Nothing happens when you don't return calls — when you don't even get calls. People write you. If it's so important that they truly need you — which would be the only reason for their calling you — nothing happens. They wait.

We have a short-wave radio for emergencies. And we can drive the six miles up to Yaak, if need be, and stand at one of the pay phones or sit on the stump that's inside both open-doored booths, and if the lines are not down — which they often are — every time a car or truck drives across the Yaak River bridge, you can hear it several miles away as you try to listen to your terrible, tremulous connection to New York, to Singapore.

Things can reach us by Express Mail in less than forty-eight hours (what the Postal Service refers to delicately as a "two-day-service region"). If it's really urgent, and the phone in Yaak is down, I can keep on driving another forty-five minutes, to Troy, population 900, where they've got three or four pay phones, or even to Libby, an hour away.

The cabin is silent. The dogs' rabies tags jingle a little whenever they get up and stretch or ask to be let out. The floor creaks whenever Elizabeth or I walk across it. The wind slaps against the rafters, the fire pops, logs sift and then fall with a muted *thunk*. If it's snowing corn snow, or if it turns to rain — one of the many shifts in the swirling wind that we get in this valley; a hard north wind changing to a hard south wind, and rain, like a seesaw battle, a tug-of-war — then you can hear that too, tapping against the glass.

At night coyotes howl, and sled dogs and Alaskan huskies all over the valley answer them, rising up and baying until all of the mountains ring with the valley's wailings, and it sounds like some village of the damned . . .

NOVEMBER 21

I'm driving across the summit, coming back from Libby. We've had two nights and days of solid snow, big flakes, falling feathers. I put the truck in four-wheel-drive low, and still it's straining. The road before me is white and unblemished, thick drifting dry soft snow, and no one's been over the summit all day. The flakes in my headlights are swirling, trying to hypnotize me. Elizabeth is waiting at home and doubtless worried, the flakes spiraling in toward my windshield, toward the center, coming straight at my heart, and I can't feel the road beneath the truck's tires. I'm following a river of snow leading all the way to my cabin, all the way home.

NOVEMBER 23

Elizabeth and I are becoming old hands at stamping our boots before coming into the utility room. It makes me laugh out loud, how professionally we do it — knocking the snow from our boots before stepping inside — how natural it is. To look at us, you'd never guess we hadn't been raised in snow country, haven't lived all our lives here.

We stamp snow from our boots exactly the same way the locals do. It makes me feel good to do it, makes me feel rooted.

NOVEMBER 24

I watch individual flakes; I peer up through the snow and see the blank infinity from which it comes; I listen to the special silence it creates.

Anything I'm guilty of is forgiven when the snow falls. I feel powerful. In cities I feel weak and wasted away, but out in the field, in snow, I am like an animal — not in control of my emotions, my happiness and furies, but in charge of loving the snow, standing with my arms spread out, as if calling it down, the way it shifts and sweeps past in slants and furies of its own, the way it erases things until it is neither day nor night — that kind of light all through the day — dusk, several hours early, and lingering, lingering forever.

I am never going to grow old. The more that comes down, the richer I am.

It is all coming down. It is all going to land on my earth. None of it is going to escape me.

For a little while this afternoon I stood out in the field, in a T-shirt, and let the snow flood down.

NOVEMBER 25

The day before Thanksgiving. Windows frosted over with root crystals of ice. I've gained fifteen pounds, about half of it muscle, but half not.

More snow. I haven't written a letter, haven't so much as considered a story. Sometimes you just drop out up here.

Dave told me that this happens. No matter what needs doing, no matter what obligations or ambitions you have, you just slow down. You sleep late. You sometimes eat four big meals a day. You listen to ravens, and keep the fire going.

I stand outside in the snow for long periods of time, in the middle of it, looking out: I cannot believe I am so rich, getting all this snow. It's all mine, all ours.

Everything's so quiet. It's more like an afterlife. I never dreamed I would live in a hard country away from people, with such quietness.

Clerks were rude to me on my way up here, all the way to Libby. I couldn't understand it. I thought of checking in the mirror, to see if something about me was particularly insulting. They were rude even in Provo, Utah, where I was at least able to count on the Mormons to be courteous, if not sometimes slightly addled, in a sweet sort of way.

But no. Something's going on out there, settling like ash over everyone.

But not me!

"Liars! Masqueradoes! Doomed pretenders!" Barry Hannah roars at his audience. To which I add, "Americanos!"

So many piss-ants out there. Why? Cheating has become a value, an accepted norm; it's in the culture. A man sold me a battery-operated portable TV (American made; an advantage, I thought) and assured me it would get good reception in the mountains. Hell, I bought it in the mountains, in Santa Fe. Elizabeth was driving, and I wanted to watch the OU–Nebraska game, and all the other football and basketball games, into the winter, on our long trips in the truck. It seemed like a great idea. The bastard had the thing on display, hooked up to an aerial. He said I'd get the same reception without one, but I didn't get squat. I looked him in the eye when I asked him that, if it would get the same reception, and it was his stinking job to lie to me. It didn't bother him; he'd stopped thinking about it, or about the consequences, long, long ago.

He was about forty, maybe a failed actor, and moderately handsome, with a cheery way about him, a sort of smirk on his face, as if you and he were the only ones smart enough to buy one of these little televisions on sale.

After I got home I found out that someone in Alabama had leased a piece of land to my partner in Houston and me for oil and gas for $2,000. We'd paid him; he'd said he owned the mineral rights but he didn't, and he won't pay us back. I think he knew he didn't own those minerals; he knows now, for sure. But he won't give the money back.

Ice crystals are melting from the greenhouse window, but my hard heart stays the same, demanding. Coals are popping and banging inside the furnace, inside the wood stove: such roaring heat. This froth-without-vent will kill me someday if I ever slow down, if I ever stop exercising. Such anger is not a good habit, but I can do nothing about it except fight, or run.

So I run. Thinking about the expatriate writers of the twenties and thirties, and the most famous one — was it Ezra Pound? — who said he had abandoned his country because he could not bear to see what it had become.

How would he like it fifty years later?

I'm hiding up here — no question about it.

The decay in our nation is frustrating. We truly are becoming senile. I feel as if we are very near the end; each time I go to a city I feel it more and more. All I want to do is get back to Yaak, back to the snow, back up into the mountains.

I'm wondering if I've already fizzled out, died, and up here, in the snow and the mountains, I have already begun an afterlife. I think that is what it may be. I have never seen any of these things before.

NOVEMBER 26

Thanksgiving. The Texans up here — a batch of us, almost one third of the valley — gather over at Bill and Carol's cabin for dinner. Big flakes coming down, and it's odd not to have a football game going. R.C. is the oldest here, among the woolly, transplanted, long-ago Texans. He says he used to be an electrician in the Navy, and he's slightly loony, I think, and a certified genius. He's got black horn-rimmed glasses, through which he's always squinting at everything, squinting at you when you speak to him, studying you harder than you've ever been studied. It's as if he's just awakened from a long nap. R.C. was one of only three men in his whole troop, or whatever they're called in the Navy, who knew Morse code — and R.C. says the other two knew it only because he had taught it to them.

"Man, you got to be fast," says R.C.

I see his fingers tap sometimes when he's talking; sometimes, as he's talking, he still drums out on the table what he is saying. It's as if he is speaking to you, but also, at the same time, speaking to someone more important than you, someone out there in back of you. This was all over twenty years ago in Vietnam — R.C.'s sixty now — but I can see his fingers still tap it out, whatever he's saying, or thinking.

Maybe you remember the famous message sent from some general over in the States, across the water via Morse code, going from battleship to battleship, humming through the air, through the night, across the sea, over to Southeast Asia: CAUTION — REPEAT, CAUTION — THE SOUTHEAST COMBATANTS ARE ATTEMPTING TO SUBVERT OUR FIGHTING FORCES BY SUPPLYING DRUGS TO OUR FORCES — REPEAT, CAUTION — and this message skipping from boat to boat until it reached the one boat that caught it and turned it around and tapped it right back — SEND MORE DRUGS. It was me, R.C. says, who sent it.

The funny thing is, of course, that R.C. says he despises drugs, abhors them, has never even tried them. He's as pure as ice. He's all dots, not even any dashes.

We're waiting on the turkey. The first one was spoiled, so we had to roast another one. It's going to take four hours. It's late afternoon and we're only just now beginning to smell the second turkey. Bill is talking about a trip through Africa that he and Carol and Fraecke and Della took about ten years ago in VW buses. He's showing us the photo albums — pictures of the Nile, pictures of elephants.

"One night we camped in our buses on a plateau in the desert, up above the river," Bill says. "I woke up in the middle of the night — I think it was the moon that woke me — and I looked outside, and there were all these hippos out in this desert, standing around our vans. They had us surrounded. Each one of them was as long as the bus itself. I think they thought it was another hippo, two other hippos, and they were inspecting things. I didn't know what to do. I didn't know they ever came out of the river."

"They come out at night," says Carol. "In the daytime, the sun blisters their hide, so they have to stay under water. But at night, when it gets cool, they come out into the desert like regular things. You can see their hoof prints in the sand."

I see R.C. scowl, and he studies Carol so intently that I

feel she might shatter. But she's smiling sweetly, it's the truth. We pass the scrapbook around, look at more pictures. I look outside and deer are walking through the back yard in the falling snow, pausing to paw through the deep snow with their hooves, like horses, searching for something to eat. It's cold out. One of them, a big buck with antlers, comes up to the window and looks in. His breath fogs the window glass. He's got large antlers. The cat jumps down from the rocker and runs into the other room, frightened.

Pecan pie.

Sausage dressing.

Sweet potatoes.

The garden's under two feet of snow now, but that's where the potatoes came from earlier in the year. We all live miles from each other, scattered throughout the valley. Whenever one of us wants to get in touch with the other — something urgent, say — we can call up on the CB radios we keep in our kitchens and in our trucks. Say Fraecke's truck gets stuck in the snow while he's out cutting wood. Suppose it's dusk and zero degrees and the temperature's falling fast and Fraecke is twenty miles from home and he's got a bad leg (okay, a logging accident) and can't walk it. So he gets on the radio, and in my kitchen, baking bread, Elizabeth hears him, and she gets on the radio and tells Fraecke I'll be right there — or maybe crafty R.C. will cut in and say he's already headed out that way — or perhaps R.C. will stop by and we'll both go out there together to pull Fraecke out, so that he can get home.

We try not to abuse it, using these radios to talk to each other. The range of the radios is about twenty miles. We use a channel that no one else bothers to use, a station that no one else can hear. But we don't abuse it — we don't just call up to chat and talk back and forth. That's cheap stuff and not why we are up here. When we want to talk to someone, we go over to that person's cabin and go in and sit down and stay the afternoon, the evening sometimes. When we want silence, we stay home. But we stay there listening to it on the radio always, always keep the silence turned on, in case one of us gets in trouble.

"I've seen it lightning during a snowstorm," says R.C., looking at Carol, still trying to challenge her hippo story, perhaps. "I've seen a lot of funny things. Any of you ever see bolts of lightning coming down through the snowflakes?"

None of us has. I think we can't even imagine it.

"I've seen it up here once," says R.C.

Carol checks the turkey. Hugh, from Kilgore, gets up and cuts the pecan pie, begins bringing us all pieces of it on tiny plates, with dessert forks and napkins and coffee.

We'll eat dessert first. We'll wait for the turkey to cook.

I see R.C. is watching the snow out in the yard, watching the gloom of winter's end-of-day come sliding in. More deer are out in the yard, perhaps twenty of them. The wood stove groans and purrs, pops. There's still heavy snow coming down, and I know when I step outside tonight how quiet it's going to sound.

I see R.C.'s watching the snow almost appreciatively, holding his pie in his hand and eating it like that instead of using a fork.

"My parents used to call me *mental*," he says, still watching the snow.

But there aren't any of us up here who don't feel that way. We've all had run-ins with society, with crowds. We've all always been able to hear something better, something different, whizzing just above our heads — a sort of a buzzing feeling in the back of our minds, sort of a crackle — and for a long time we did not know what it was, weren't sure what was going on. It was always up there, though, just beyond us, just always out of reach.

Sometimes, along the Yaak River in the summer, at night, we can park right by the water's edge and pick up what they call "skips" on the CB radio — not the usual ground waves but high-altitude waves that oscillate along, following the upper, curved rim of the atmosphere — traveling faster, traveling higher, coming from long distances — Nebraska, Florida — radio waves that get caught in the bowl of our valley, finally, up in the high mountains, and that ricochet around them like mad hornets, and sometimes in the summer, late at night, we can pick up the CB conversations of farmers in Kansas, of people in Oklahoma, two thousand miles away, their voices as clear as a bell. They think we're bullshitting them, these people so far away, when we tell them we're high in a valley in Montana. They think we're just around the corner, in downtown Oklahoma City, watching them and laughing at them.

"Tell them about skip waves, R.C.," I say, and R.C. keys the mike and tries to explain it to them earnestly. But I can tell they have no time for it. We can tell they're not listening.

The turkey is ready. The hot knife cuts through it and the pieces of meat fall away. We pour gravy over the dressing, over the turkey, which is moist. Elizabeth made the rolls, and they are steaming. We slide butter into them. Della made the butter. The meal is so good that it makes me shaky; I feel like weeping. I want it always to be like this. I'm sitting closest to the stove and my feet are warm. The coffee is good. Carol brings us all another piece of pie, with homemade ice cream on top of it. Fraecke gets up and sticks his head out the door.

"It's still snowing," he says.

"Nineteen fifty-six Buick Sentry," R.C. says, and some of the guys who know about such things make appreciative groans. "Strongest production car ever made. Should have been outlawed. Cost me twenty dollars once."

There's no need to ask R.C. why. I'm so glad he's here.

"Friend of mine sat me down inside one once, put a twenty-dollar bill on the dashboard, and told me he was going to accelerate from zero to fifty, and that if I could reach up from the front seat and grab that bill before we got to fifty, I could have it. But if I couldn't, then I had to give him one of my own. So he hits the accelerator and I get thrown back in that seat, and I'm trying to reach up and grab it, but I just can't," R.C. says. "That thing was pulling two Gs. It was the weirdest fucking thing. I could barely move my hands — it was like they were moving in slow motion. I just couldn't get my hand up there in time. It was just too much force."

One more thing about R.C.: he's keeping a little buck deer this year, a baby yearling — just a fawn, really — that he found. It had been wounded by hunters. R.C. caught it somehow and took it seventy miles, to a town, to a vet, and had it doctored, and now he's got it back at his place, keeps it in the living room as it gets better, and is feeding it, healing it, changing the bandages. He says he spends two or three hours a day working with the little thing.

DECEMBER 3

Snowy skies upon awakening. An inch fell in the night as we slept, unknowing. But now the temperature's all backwards, going the wrong way, climbing back up the scale. A chinook is blowing warm, moist, irritating air through the skinny valley, chasing our winter (which I'd only recently started to feel we'd earned) away. I'm cross, as a result — snow melting, sliding off barn roofs and cabins, and icicles dripping. I feel as if something's being taken from me.

And my friend Kirby, who was visiting, is gone. We're alone again just when we had gotten used to socializing. It's disorienting: are we lone wolves, or are we herd creatures — deer, elk, even sheep? My eyelid's fluttering wildly. We miss Kirby. And this damn chinook — we *want* it to be cold, we *want* the snow, desire the isolation, desolation, insulation, silence. It's why we're up here, it's why everyone's up here. Not a particularly flattering thing to admit, but we're all on the run from something, and it makes us feel safe, this isolation, but with the snow melting and the roads passable again — accessible to any damn tourist who wants to visit, any old pilgrim — it makes us feel *exposed*.

I want it to get colder, harder, deeper — not warmer.

It's all wrong today. I'm grouchy. Elizabeth and I both are. It's reverse cabin fever. You won't read about it in any medical textbooks.

I want more snow. I want to be snowbound. And I'm confused too. We had such fun when Kirby was here.

I'm supposed to be a hermit, but what a half-assed hermit I'm turning into: running away to the woods in order to discover that I love people, friends.

But I love the woods too. They're right outside the window. I'm touching my fingers to the cold window panes of the greenhouse. There are wild things just beyond, in the woods. Those wild things — and they are watching me, though I cannot see them, though I see their tracks in the snow — understand my fluttering eyelid, my fluttering heart.

DECEMBER 9

We get a lot of preachers up here: missionaries, young ones. I think our valley is a proving ground, a test track, before they go up to the Yukon. I've been waiting for the ubiquitous Mormon missionaries to surface: strong, mellow, easygoing, confident, a perfectly matched duo — training! training! — riding those bikes all over the place fleshes out their arms, gives a shine to their cheeks. They can take on the world, they know about the world and are not on the run from it, as were the two revolutionaries who came by yesterday. The missionaries simply believe they've seen something better, that they *have* something better, and are merely burdened by guilt, having it that good, while the rest of us know nothing about it. I think that's how the Mormons must feel, how they go about their business.

But the man and woman who were by here yesterday drove up around eleven o'clock in a new, innocuous Ford van: two-tone, studded snow tires (but still not much clearance; they *had* to be local, I thought), and with their children bouncing around, percolating inside (five? six?). I thought they'd come to look at the place: potential buyers. I was prepared to tell them all the pitfalls and about how severe the winters are, and that if they wanted to buy the ranch, well, they needed to understand that all that wood stacked outside didn't go with it.

The little man stood at the doorway smiling oddly, like a beaten pup hoping not to be beaten once more. The woman — tall, well dressed, almost elegant, black-haired, a beauty — stood meekly behind him and to the side, feet and knees touching, so straight was she standing, clutching a little purse, as if waiting to be asked in — but watching the little man. My first thought after deciding they weren't buyers was that they needed help. They looked lost, or worse.

The first thing the man said was, "Do you belong to any particular religion?"

This guy was dangerous.

I told him that I was not of any particular religion. So tired of churches I could roar, is what I wanted to say. But I could tell, too, that this would only confound him, that he was out in left field. Like the Mormons, he had some new idea, but unlike them, he had not test-driven it yet. It occurs to me only now that all the children with them might have been kidnapped. He was too tiny and haggard to have loin-sprung that many, and she was still too young, too erect, and too vague-looking to have been a mother that many times. It's possible that, like the children, she was brainwashed. I suffered through a few random dramatic readings (they'd been hiding their Bibles and pulled them out from behind their backs). The man was getting excited, finding passages where various lesser players — Titus, Philemon, and such — went around asking if faith was enough and saying that it seemed to *them* that faith without works was *dead*.

The little man was getting so jumpy that he began skipping lines and mispronouncing words, but he sure could rattle off the numbers of the verses, even if he didn't know what they meant.

Sensing my impatience, even my anger, perhaps —

surely he had developed an acute sense for such things — he flushed, like a bird breaking from cover, and went on showing me these key passages deep in the heart of the Bible, phrases having to do with the weather, with the life habits of fish, with all sorts of trivia, and he started listing other verses he had ready to back him up. "Lessee, lessee, in Psalms 32:17 . . . is it 17? No no yes yes in Psalms 32:17 —"

"That's right," the woman said, leaning forward slightly to deliver her first words.

Then the man was speaking again: "It says, it says —"

I had to cut the little man off, I just couldn't take it, couldn't be polite anymore. I'd given him five minutes of politeness as he read, but he was taking advantage of me. I told him I was typing, and I had to get back to work.

The woman put on her embarrassed, I've-been-slapped look, and the little man's face started to twist into a sneer — a good, old-fashioned Paul Lynde sneer. "Oh, I didn't know you were *typing*," he said. He left some tracts with me — children's coloring books, actually — about how in the Last Days a new government should be set up, a government run by the meek (this guy would have been president). They hurriedly got back in the van, turned around in the snow, and drove back down the driveway.

I saw them later in the day as I was going to cut wood. All the way across the valley, their new cream-and-brown van was parked outside a cabin on the far side of the river. I could make out the small figures in the snow, climbing back into the van. Later in the afternoon I saw them leaving the mercantile, looking beaten. I waved, then laughed, thinking of that woman in Yaak, the one who yelled at her husband ("Hey, Numbnuts! Some missionaries here to see you!"), giving them the boot, a boot to the meek. I wondered if they had swung through the Yaak on a whim. I can see them studying their map, planning their route, pointing to the little intersection, the odd word: "Yaak. That sounds like a barbaric place, dear, a place in need of help." They had better jettison some of that meekness and shore themselves up, I thought, if they're going up to British Columbia.

There must be some guide book, some mail-order correspondence course, that directs so many of them through these parts. There aren't many souls to save here, of course, but perhaps that's the way to get an "in." People are so glad to have a visitor of any kind, especially in winter, that they'll actually listen, in order to prolong the visit, rather than run them off rudely, like in the city.

Perhaps all the snow in the world will fall, burying everything, such silence, and then I will come out of it in the spring, different, cleaner, not born again so much as built up. I'll laugh at more things, and not get so angry at decadence, at laziness, at deceit and the theft of time, the theft of truth, starting with the President and going all the way down to the grocery store.

Snow's coming down, a hard wind out of the north. The generator's got a short in it, and it's hard to start. I gassed up the backup generator yesterday and made sure it was working okay. One of the large hanging pulleys in the barn is swinging wildly, making a squealing, almost mocking sound in the wind, a sort of yes-no, yes-no, yes-no, squeak, so loud it can be heard over the high wind. I could go to the barn and climb up and oil it. But it's really the only other thing out there, between here and the woods. It's a grating sound, not normally pleasant, but I almost like it this morning.

*   *   *

Ann ran away today, delirious in the falling snow — leaping and snapping at the flakes, stopping to rear up and dance on her hind legs. I tracked her to the neighbors', where she was sniffing around, trying to get into their rabbit hutch. Anybody else would have . . .

It's scary. I will watch her. Nothing will happen to her.

A heavy rainstorm tonight. The snow turned to rain, oddly, at dusk.

I've been thinking about that "mysterious" hole in the earth's atmosphere, right over the North Pole. Can we lose ourselves to space? Can all of our gravity be drawn up through that hole, and us along with it, into a sort of vacuum? I try to picture all these people flying through the air, headed for the North Pole, Revelations, and the Second Coming.

And Ann, left behind, hunting rabbits in the snow.

Already at dusk, trees were beginning to break and fall across the road, loaded down with ice and snow. It's the ice that gets them. I had forgotten what someone told me back in the fall, to always carry an ax in your truck, so that when there's a tree across the road, you can still get through.

I suppose theoretically a mistake like that — forgetting — could kill you. You could get stranded up on the summit. The temperature could plunge. You might not have enough gas to keep the truck running, the heater going. You might not have your down sleeping bag in the truck with you.

You could stay in your icebox of a truck and freeze. Or you could try to walk it, get lost in the snow, and freeze. I'll remember to carry an ax, after seeing what this storm does: trees down everywhere.

*   *   *

Yesterday I went down to the horse pond with an ax and cut big squares of ice for the freezer, to keep my meat cold (the propane freezer's exhaust pipe is plugged up, needs cleaning). When I lifted the blocks of ice out of the pond, little water-bug crustaceans (water striders, water boatmen, doodlebugs) flipped around sluggishly on the water's surface, suddenly out in the open. I couldn't believe they were living under all that ice. I put some of the ice blocks back in the pond, trying to cover the insects up, but don't know if it worked or not. It was pretty amazing to see them crawling around on that ice, some of them even burrowed into it, on the underside.

Dave Pruder came by this morning to feed the two horses, Buck (the line-backed dun; friendly) and Fuel (dark brown, almost chocolate; wild, shy, and beautiful). I told him we'd be glad to do it on the days we were here.

It was fun, carrying armloads of sweet hay from the barn across the road and into the pasture in the cold weather. The horses came running, their hooves making a solid sound on the frozen earth (Fuel hung back a little). We gave them both a big can of oats too.

Soon, maybe tomorrow, we are going to build a shelter for the horses. Dave hadn't counted on all this cold rain. They can take all the snow in the world, with their furry Montana coats, but the rain's a bad thing, rain turning to ice. It's pounding against the windows, blowing in gusts. I hope to get over to Seattle and Vancouver later this winter. Money's still a worry, though, as always.

But I'm very dry, very warm. Are the horses, out there in the cold blowing rain, looking up at the lights of my cabin as I write, wondering why they have to be that cold, that miserable? Surely they're down in the trees, out of the weather. Out of most of it, anyway. I hope.

We've got a barn, a huge one, down by the road on our side of the pasture. Dave should at least bring them into the barn. I've been up here a little over three months now, and I'm telling Dave his business. But he should. It's dry in the barn, and warm.

I bought Elizabeth a newspaper today, the *Missoulian*. She sat in front of the fire and read it through twice. I'll have to keep an eye on her, take care of her. We'll have to take care of each other. There's nothing but two million acres of woods, and then Canada.

Me, I love it. I may stay up all night, reading and putting logs on the fire and listening to the rain.

December 10

I go through the candles so quickly. Down to three nubs this morning.

The wind hurls corn snow against the windows; it can't make up its mind which direction to take, and is tapping first against one window and then the other. A few grasses still poking through the snow, blown flat when the wind gusts, bending first one direction and then the other.

A missile treaty was signed two days ago; I read about it in yesterday's paper. Thinking now about the Soviet Union's natural resources: mega-giant oil fields, oil reserves, minerals, and forests up in Siberia. All that room, I'd like to explore it on skis, maybe with some sled dogs, someday.

For an early Christmas, my brother Frank, a reporter (he won a Pulitzer two years ago), has sent me a photojournalist's canvas carryall, one that goes over your shoulder — one of those Banana Republic things. He wants to save enough money to fly down to Central America, buy an old car, and drive around — to do who knows what. Yesterday I read that the Sandinistas captured a fifty-seven-year-old Iowa farmer who was flying around down there in a tiny Cessna. Fifty-seven? Somebody's Iowa granddaddy, captured by the Sandinistas.

Because Frank is my younger brother, I've told him I'll try to go with him — might be a good short story in it —but I would rather go to Siberia, or even up to Alaska. Or here — maybe I could stay here.

Maybe I can find a huge news story for him to report on up here, instead. Maybe I can make one up, or even create one. I'll go into the garage and build a mock nuclear warhead. I'll use hollow aluminum tubing and put fins on it, paint *USSR* down the side, put a ticking alarm clock inside it, carry it up to Hensley Mountain, and stick it in the snow so some hunter will find it. Maybe I'll make twenty or thirty of them, plant them all around Yaak. Climb up on the roof of the Dirty Shame and stick one in the chimney. Then I'll make a phone call.

"Hey, Frank," I'll whisper, "get up here quick. Hurry."

I'm not wild to go down to Central America. I've no desire to see Lebanon either, or even Belfast. I like Montana.

I suppose that makes me a pacifist. Pacifist of the Pacific. And passive: just holed up, letting it snow. I don't even write letters to the editor anymore (well, not many).

I'm riled now, thinking of *USA Today*.

"Kittens: Cats of Tomorrow?"

Or this headline: "Exciting Murder in N.H. Has All the Elements of Made-for-TV Movie or Miniseries."

Masqueradoes! Doomed pretenders!

DECEMBER 12

The sky is so blue this morning it's like a different land; I was truly surprised to see it after so many days and nights of snow and rain. There's a good hard twenty-degree cold, and the greenhouse windows are latticed with snowflake root-crystals that are losing their magic, dissolving now in the heat of my just-started fire, even before this sentence is finished.

The good thing about writing in a small room is that it gets warm easily.

Tore up my knee again two days ago, climbing around in a slash pile at dusk, looking for log ends to cut with the snow falling. I slipped through a hole in the slash and fell forward with my ankle jammed between two logs. I banged my knee on the side of a big log end and roared, but held on to the saw, which was still running, and pulled free, climbed down, limped around, then went back to sawing. Yesterday I couldn't even walk on it, thought I was going to need it operated on — more downtime, spend more money. But this morning I can flex it, stand on it. It still hurts, but I can move it, and it's going to get better.

My knee and this hard clean blue sky: these are surprises. Have to run into town to buy some points for the generator,

make some copies, do some mailing, buy some groceries —
will try to make it back before dark, to go cut the Christmas
tree. We bought popcorn earlier in the week, for orna-
ments.

I'd also like to get a wood run in today.

Good coffee. A good December day.

DECEMBER 13

It got down to zero degrees last night, but we were warm
under many blankets.

Went outside to get firewood this morning. I had just
washed my hands and put my contacts in, and reached for
the doorknob to go outside. I guess my hand was still damp
and it stuck to the metal, wouldn't let go. When I pulled it
away, I heard that Velcro sound. Scared me worse than
anything. Just ice crystals was all, this time — not flesh, not
at zero degrees. Then, as I was walking out to the greenhouse, carrying a little cup of milk to go with my coffee
(mama's boy), a one-hundred-yard walk, the milk froze.

Zero. Feels good. I'm going to cut a Christmas tree today — didn't get back in time yesterday. Also some more
firewood.

DECEMBER 15

I met someone last night at the Dirty Shame during *Monday
Night Football*. After we had exchanged enough comments
about various plays, in which we both proved our intelli-
gence and knowledge of the game, he felt it was okay to
volunteer his name, Joe. I gave him mine. Handshake.

Joe is the brother of a friend of mine in the valley, he tells
me in one breath, and in the next lets me know that another

of his brothers, who used to live up here, is in jail now. Joe seems relieved about that. I think this other brother must be the guy whom Roger Craig, the preacher, feared so much: the one who was in jail, then out, and is now back in again, for allegedly stealing a certain cruise boat and taking it to Hawaii.

Dave's burritos were good, meatier than ever and bulging, as large as the dinner plate he served them on. Usually I eat two, but I was beaten down from sawing and splitting all afternoon, so I had three. My file grinder broke, so I had to attempt sharpening by hand, but as I'm right-handed, I overcompensated on the down-sloping teeth. I found out later that everyone does (there isn't a logger in the woods who can file too many times before getting the teeth all crooked on one side or the other). As a result the saw was binding, and also cutting on a slant, so that I was cutting about one and a half times as much wood, and having to cut with my back, really sawing, as if the machine were a crosscut saw and not gas-powered. To make things worse, I had not been rotating the bar every two or three cords, the way you're supposed to. It made great sense when it was explained to me, but there had been nothing about it at all in the sixty-page instruction booklet, so I didn't know any better. The groove in the bar around which the chain spins as it does its cutting gets worn down at the bottom when the chain binds. Eventually, with a dull chain, the bar will get worn so unevenly — like tires out of alignment — that you'll always cut on a slant, unless you rotate the bar to distribute the wear, the friction. Plus, if you're right-handed, you'll bear down harder on the right, sending the blade out to the left; your logs, if you're able to cut through them at all, will begin to look skewed, with a worn bar and a dull, uneven chain — as did all of mine yesterday.

Dick McGary explained all of these things to me last night at halftime, for who knows what reason — invaluable philanthropy — perhaps it was just the effect of a few beers, or the onset of winter proper.

He was puzzled by my saw at first when he sharpened it on his machine (for $3), because the chain brake locked on him. The brake is a safety device that's supposed to hit your fist if the saw kicks back violently, and pushes a button that kills the engine, locking the chain up, so you don't do damage to your head. But Dick said that all the loggers take the brake off their saws, viewing it as cumbersome. After having spent thirty or forty years in the woods with a saw, they simply know how to handle one. I once watched a logger up on Hensley trimming limbs with a big pro's saw, an old, beat-looking thing. This logger was in his late fifties at least, and sixty pounds overweight, but he was catwalking nimbly along a big log, wielding the buzzing saw like a baton twirler, like a ninja, snipping off arm-size branches as if he were swatting flies — backhand, sideways, jabbing at them with the nose of the saw (usually a *sure* way to get kickbacks). To me, it seems that taking the chain brake off a saw, even if a pro does it, is like a race car driver not wearing a seat belt.

I bought 2,270 feet of buckskin larch from Breitenstein yesterday for $120. He had several great larches felled on his property — some of them over three feet in diameter — trees that had been injured in the great fire of 1910, and again in 1931, and had only recently begun to die. Green larch isn't any good, it won't burn, but dry buckskin larch is almost atomic in the heat it throws out.

You can find all the small lodgepole and fir you want in the slash piles at logging sites — wood I try not to feel too guilty about burning, since the lumbermen torch their slash piles when they're finished anyway. It's going to go up in smoke regardless; at least I can keep it from being wasted.

But the larches — trees that were big even in 1910, when the fire came through and burned off all the bark, turning the giant trees an ashy gray color (hence the name "buckskin") — those are the gems for firewood.

Breitenstein had felled some of these dead giants — they were starting to fall over in his neat woods — and scaled them (trimmed all the limbs off) and marked them with blue crayon on the butt ends. The largest log was eight hundred board feet (one thousand board feet equals almost two cords). This huge log was the shortest he had, but also the most massive: with my crooked saw, yesterday afternoon I was able to muscle through twelve cuttings of it, plus twelve of a smaller, younger log. I think I still have about six or seven more cuts to make off the big guy, and then another one almost like him, plus two more smaller ones — about five cords in all.

I'm going to buy one of the largest larches from Breitenstein someday. He's got them felled all around his forest, hidden like giant candy canes in the waist-high ferns in summer, and in the snow in winter. The ones you buy he drags, or skids, out to the road with his dozer and prunes all the branches off, so all you have to do is move down the line, cutting.

As I've said, there aren't very many great larches left. The European larch (*Larix decidua*) is like the giant redwood; its own greatness is what's killing it. It's the tree everyone wants: it's heavy, dense, yet splits easily — pickle larch, they call the great dead skeletons, because, though a strong man cannot lift even one of the logs, a boy or a small woman can split one easily, with a single blow of the maul, if it hits a seam, visible as either a dark streak or the tiniest of cracks, a fissure, leading into the heart. Before they're all gone, I'm going to buy one of those great felled trees, trailer it home to south Texas, and put it on my parents' farm, to show my family, and my children someday, what firewood can really be: how hot it can get, how long it can burn. We'll crank up the chain saw and cut off one stump per year, at Christmas; the burning of the Christmas larch, to remember what a forest used to be like: a magical land of giants.

Am I a plunderer, a taker, for going into the woods on a snowy evening and exchanging money with Breitenstein for the soul and body of one of these trees he's already felled, trees that are dead but still magnificent, and that will never be again? I don't know. What I do know is that he sells them for the same price to a company over in Idaho, but he likes to offer them to the locals in the valley first. I know that the Idaho company has told him they will buy all he can get for them, no limit; and that he spends all winter out in the woods, working his dozer, running his saw, felling the dead giants, his hobby. Perhaps he has no belief in immortality. The trees would rot anyway, would become spongy and useless (except as nutrients, of which there are plenty in these Pacific Northwest rain forests) if neither I nor the Idaho company bought them.

There is another eight-hundred-board-foot log, this one with a crooked blue "Don T." scrawled on the butt end. Don Terlinde — who lives up the road, and who has perhaps even more wood than I do already stored away for this winter, for future winters — bought it, has it sitting on two boulders, and might not get around to cutting it until summer, but he, too, wanted to buy it ahead of time, while it was still around.

After the larch is gone, it is quite possible that the people in this valley will be cold again. They will have to move into smaller cabins, which they can still heat with lodgepole and fir. They might — like the mercantile will next year, like the Dirty Shame has done already — change over to propane, the fuel of affluence, an exchange of money for heat, instead of labor for heat.

I am going to save some larch for my children. Maybe not for everyone's children, but for mine. And for my best friend Kirby's son or daughter, due in May. Perhaps this spring I'll drive into Houston pulling a sixty-foot larch tree behind me on a gooseneck trailer and leave it in their front yard for the child to grow up with, to play on, to shoot arrows into and whittle names in. I got a card from Kirby and Tricia, asking if I would be the godfather, saying that they very much wanted me to be.

I know what immortality is. I am beginning to understand it.

Yesterday afternoon, going into Yaak to take a break from sawing and to have Dick look at the saw, I noticed Mary Breitenstein riding down the road on her big bay horse, leading a muddy-legged, wild-eyed, cream-colored colt behind her; it was obvious that the colt had fallen through some ice into deep water. When I stopped, Mary said yes, that was what had happened, that coyotes had chased it onto the ice and she'd had to pull it out. Mary, who is twenty-four, looked worse for the wear than the colt did — she said she was walking the colt up and down the road to keep it from freezing — and I waved and drove on. The heater in my truck was running, and it felt good.

Then later, at night — as we were driving into Yaak for the football game, the sun down a good two hours — Elizabeth and I saw in our headlights a woman riding into town, coming down off the snowy Eureka road on a dark horse, all bundled up. Cold vapor frothed from the big horse's nostrils; the woman, whom we'd never seen before, was headed for the mercantile before it closed at eight. Snowflakes were coming down like feathers.

DECEMBER 16

It's not like we're total hermits, only that most of us want, as Thoreau said, to examine our lives, as well as the world we live in — a world that, up here, is not controlled by others as much as by, believe it or not, one's self.

We have dinner parties every few months, and invariably, late in the evening, after much wine, someone will put the valley's favorite song, "Night Rider's Lament," on the turntable (if the generator's running) and we'll sing along with it, over and over.

> While I was out a ridin'
> A graveyard shift, midnight 'til dawn
> The moon was as bright as a readin' light
> For a letter from an old friend back home
> And he asked me
> Why do you ride for your money
> Tell me why do you rope for short pay
> You ain't gettin' nowhere
> And you're losin' your share
> Boy you must have gone crazy out there.

We've all — all except for Breitenstein, who doesn't go to dinner parties — come out here from someplace else. We all get such letters from family, and even friends, saying: It's too cold ... But you need a telephone ... How do you get by without electricity ... What about the animals . . . What about the wild animals . . . You're losing your share . . . You're missing out . . .

About "Night Rider's Lament," about moving away from things in order to examine other things, Jim Harrison says it well too: "The woods can be a bit strange. It takes a long time to feel you belong there and then you never again really belong in town."

I'm discovering things about myself up here, things I should have known by now, but don't. I don't think this reveals a complex personality so much as a simple mind, but it's the only one I've got, and I'm glad to have it out here, away from town.

I used to think it was bad, a failing, that I had to be in the wilderness to be happy — away from most things. Now I'm starting to discover that's irrelevant — whether it's good or bad, a failing or a strength: totally irrelevant. It's just the way I am.

Water, on this side of the Divide, flows to the Pacific.

DECEMBER 18

Victor, Dick McGary's son, regales me with chain saw injury stories each time I buy groceries at the mercantile. He was splitting wood once, on a cold day, and he swung at a green stump and had the maul bounce right back off the stump and hit him in the face, knocking out some teeth (no one up here has a full set of teeth). Sometimes when that happens the maul hits someone in the forehead instead of the teeth, and kills him.

Victor shows me little bumps in his hands from where the maul has thrown back slivers of metal (this happens when it gets worn down, flattens out, and loses its edge) that have lodged under his skin. Bumps and slivers of metal: he shifts them around, making them move, describing to me what the morning or afternoon was like on each of the days a sliver hit him. He gets a magnet off the counter and passes it over one spot in his arm, and the bumps, like goldfish in an aquarium, rise, congregate, follow the magnet.

These are accidents, excusable in Dick's eyes. Dumb things, like cutting your thigh open or putting the chain on backwards and then complaining that the saw's not cutting — Dick has less tolerance for these kinds of stories. The deep and ragged, sometimes fatal thigh cut is the most common saw injury. Some of the loggers these days wear Kevlar pants — made of bulletproof plastic, they're like chaps or snakebite boots, only they come up over the thighs — but other loggers claim that, as with the chain brakes, the pants slow them down. They say that when they are felling trees, the thing that saves their life is their ability to run away from falling limbs, get out of the way of falling trees, and that they can't move in the Kevlar pants.

Dick interrupts himself to show me another stupid thing. He can barely bring himself to talk about it, he's so disturbed by it, and is able only to point a thumb at it, tilt his head toward the offender in a jerky, half-grimace acknowledgment.

It's a battery sitting by his battery charger in the corner of the mercantile, but it's not hooked up to the charger. The reason is — he makes me take the battery caps off because he can't bear to look at it again himself — that inside, the cells that hold the distilled water are frozen solid.

"The guy," Dick says, and his voice is thick with disbelief; he can barely speak he's so choked up; "has a college education, and wants to know why it's not working, wants to know if I can charge it." Dick's so rattled that he has to light a cigarette, and unlike most Yaakers, he rarely smokes.

I shake my head and cluck, pretending to be amazed. But yesterday, I cut a leg halfway off my overalls by accident; I had finished sawing one log and was moving to make the next cut, and the chain was still moving, though I had let up on the trigger — the way a bicycle wheel coasts for a while when you spin it, that's how the chain keeps going on a saw, even when you're not revving it — and I felt a strong tug at my overalls, looked down, and the cloth of the left leg was torn all the way around, midway up the thigh. There was also a little tear in my long underwear, which I was wearing beneath the overalls.

I checked carefully to see if the saw had gotten my leg too — I wasn't sure, because of the cold or because of the shock. Miraculously, like the luckiest of pilgrims, I was all right.

I kept cutting, and tried to remember how lucky I was.

But I was less lucky that day than Elizabeth, who, while standing next to the fireplace, trying to stay warm, caught herself on fire.

I didn't see it happen. I just drove up in my ragged overalls five minutes later and saw her standing out in the snow, not dressed very warmly, as if waiting for something. She'd had to get out of the house after it was over, she was still so frightened.

She'd been wearing a heavy flannel nightgown over her sweats, and a spark had evidently popped through the wire mesh; she hadn't had the glass windows closed all the way. At the time, Homer and Ann were sitting in the chair together, cold themselves, under a blanket. Elizabeth laughed at something Ann was doing and moved closer to the dogs (the hem of her gown on fire by this time), cooing and saying things to Ann, who was no doubt terrified by this new game, this flaming mama coming toward her with arms outstretched. At that moment Elizabeth happened to see a reflection in the front window, which she thought at first was a car driving up with its headlights on, but which she then saw was the reflection of flames behind her. She turned to see where the flames were — large and dancing by this time — and the turning motion fanned them, sent them up the outside of the gown nearly to her elbow.

Elizabeth says she slapped the flames out with her hands, and has blisters on her palms to prove it — but looking at the gown, or what's left of it, I don't see how she could have put out that many flames with her hands. She said her first thought was to run outside and roll in the snow, and I got angry at that, for her thinking of the dumbest thing to do (after I'd just come within a sixteenth of an inch of chainsawing my leg). She would never have made it, not out both doors (the heavy exterior door dead-bolted, tough to open in cold weather) and then down the steps and out into the yard; nor would she have been able to pull the gown off (small neck, with a throat collar). I guess she just did what she had to do, and it was enough.

The generator is still down. I think it needs a new starter, but am not sure. The propane freezer is running rough too. The chain saw is still cutting crooked.

The lucky, stupid, foolish pilgrims. Is the land beginning to turn on us? Is it, just now, noticing our presence, a land that admits few into it, and keeps fewer? I want to tell Elizabeth that we must move more slowly, be more alert, but it is starting to feel more like a home, and there's that danger of complacency, especially in winter.

Everything is another world up here in winter, I want to tell her. There are no rules. Anything can happen. You can't take anything for granted. You can in most places, but not in new places, and not in Yaak, ever. If it's snowing, you go out to the woodshed with a rope tied around your waist, so that if a flurry kicks up while your back is turned, and the flurry doesn't let up, then at least you can find your way back to the cabin, rather than wander around in wider and wider circles, following the crazy, damning, lost compass that is in all of us.

Two days ago a man over in Spokane slid off the road in his truck and went down an icy hill into the Clark Fork River. He kept his head and didn't try to open the door as he was sinking. He let the truck get all the way under, then rolled his window down (the cold water came pouring in), took a deep breath of air up near the roof, climbed through the window once the truck was almost full of water, and swam for the top. It was daytime; he was lucky he could tell which way was up. He burst to the surface and began swimming for shore as fast as he could. Sixty-seven years old. He made it.

Entire mountains give way while men are out hunting in winter, or while cross-country skiing. An avalanche can take you away to the quietest death imaginable — sometimes with a noise like thunder, but other times with just a soft white *slump* of a noise, a noise more of air than of anything else. The mountains can slide over and hide men beneath hundreds of tons of snow, hiding them until summer, until forever. Own two of everything, says the landlord, and Elizabeth and I are taking him up on it: two trucks, two axes, two radios, two generators, extra food, extra light bulbs, extra gloves for when one pair is wet, extra boots, extra tire chains, extra firewood, and extra luck.

Don T., cutting on his log this morning, stopped and took an hour to work on my saw — barehanded in zero degrees, snow coming down — to sharpen the chain for me and see if he could figure out why it's cutting crooked. He put a hell of an edge on the chain, taking great care and patience, showing me how when it's finally sharp, the file makes almost no noise, no friction. Don T. studied the falling snow, gazed up into the fog, as he listened to rather than watched the file. Breitenstein told me later that Don is the best saw sharpener in the Yaak. It looked so easy, and I was humbled that he'd take so much time out from his own woodcutting that I said, "Let me try, I think I've got it now." He grinned and his face split into a thousand wrinkles. Then he laughed, saying, "No, you don't *got* it," and kept filing gently, laughing, watching the snow above coming down, as if expecting someone.

But the saw still cut crooked, still got into a bind and slid hard left once the bar was all the way into the log, so Breitenstein came over, perplexed that something was being done to less than perfection in his woods, and we studied it for a long time. Finally he and Don T. decided that maybe my chain is too far gone, too narrow from so many sharpenings, too narrow for the grooved bar to follow it. So I bought a new chain ($24) and I hope that will fix it. If it doesn't, then the bar is bent or worn, and I must buy a new one of those too ($37). But, as they say, I need an extra anyway.

Breitenstein looked on incredulously as I went about my woodcutting again, straining against the bind, sawing and pulling and heaving and tugging like a man wrestling with a whale. When Don T. and Breitenstein cut, their big saws sink so easily through the wood; they merely guide the saw with one hand, showing off, holding the saw like the rudder of a gentle sailboat. Watching me was too much for Breitenstein, who whooped and shouted, "I gotta get out of here!" He ran off into the woods, hurdling fallen logs, holding both hands against his head and whooping still. "Damndest thing I ever saw," I could imagine him saying later, and I am thinking now, That's nothing, you should have seen my girlfriend on fire yesterday.

A couple of weeks ago, I read an article in the Sunday *New York Times* about the writer John Berger. When he was around sixty, he abandoned his fine urban woods to go live with peasants in the French Alps. He was quoted as saying that the only way to come in late to the game like that, the only way to earn the peasants' respect, is by trying to do the things they do, doing those things ridiculously badly, and then asking their advice, so that they become the teacher, and you the learner.

But John Berger's never met Breitenstein, and I'm not sure, after watching him run screaming into the woods, that I'd want him to, or that Berger's advice, which sounded so good to me in the paper, is especially applicable up here in the Yaak.

Thinking about Dick McGary, getting so distraught over a frozen battery, another man's dumbness. Perhaps the people who live up here believe it's like a contagion, this lack of alertness. They know how fatal it can be. I can try hard, and watch, and listen. It's all I can do for now, but it's what I do best. I must remember to watch out for Elizabeth too, however, and Homer and Ann as well.

I go to sleep hard. I'm tired in the evenings, tired in the mornings, but also stronger, healthier than I've ever been. I have almost enough energy to get everything done each day, with only one or two things left over on the list.

There will be time for trips to Vancouver, hikes behind the house, downhill and cross-country skiing, for reading in front of the fireplace — but first, I have to get us settled in, and settled in well. I guess I thought it would be easy.

DECEMBER 19

A cold front came down out of Alaska yesterday, dropping the temperature from twenty above to fifteen below in less than an hour — branches and limbs blowing from the trees, everything tumbling past, and the wind biting, ripping. The temperature kept dropping after dark, crackling cold stars, plunging, bottoming out around thirty-eight, thirty-nine below, and then the wind disappeared. It's staying that cold — the air motionless, the way they said it would be — forty below at night and warming up to ten or twelve below during the day. We leave to visit relatives in a few days, relatives in warmer country, and will turn off the water, drain the pipes, lock up. But for now we sleep in front of the fireplace, wood burning around the clock, turning the pages of books clumsily with our gloved hands.

Forty below. We're a little frightened. We're at the mercy of the cold. We hope for mercy. It's as if the brutal cold is looking for something, passing over, searching. I hope it doesn't find what it's looking for here. I want it to move on.

We hide in our sleeping bags. The fire crackles, but it doesn't seem to put out any heat. We can't get close enough to it. This cold cannot last — three days, four days, a week at the most, and we must remember that.

Blankets over the trucks and car in the garages. The generator won't turn over — can't get the valves to open, too cold.

We read, we sleep. We can't get warm. But: this won't last.

January 4

I've been busy eating, and I can feel a small roll of age against the waist of my pants when I sit down. I want to run out into the heavy falling snow and howl. I've lost my logger's hardness.

There are natural highs, and lows, in winter. You eat more. You sleep more. It is only natural to put on a little fat, but I don't like it. I want to go into winter, have its beauty and silence, and play by my rules, but it's hard. And I'm so tired at the end of a day; as soon as it gets dark I'm bone-weary, almost in a stupor. It doesn't matter whether I've been outside sawing logs or cross-country skiing all day, or just sitting inside by the window, typing and drinking tea — still I'm exhausted. I'm finally learning to savor it, just to stretch out by the fire and fall into slumber, into a sort of spinning, warm unconsciousness — all the *chores* done, or almost. I'm learning to understand and relish the sweet *low* of it, this necessary putting-off-until-tomorrow. The days are gradually, by minutes, getting longer, and soon I'll be out of it, go full bore again, put on my city ways and do the work of three men — but these short, dark days are bigger than I am, larger than the chemical stirrings going on in the back of my brain, and I've learned that if I fight it, I'll only be more tired the next day.

At noon Elizabeth is off to play pinochle at the mercantile. I've put chains on the tires, and asked her to call in on the radio when she gets there, and again when she's ready to leave. It's about six miles, but more of a trail through the woods, these days, than a road. I drove in to the pay phone at the mercantile yesterday to make a call, and Becky came outside, looking a little wild, a little giddy, swinging from one of the merc's porch columns, and she asked, then corrected it to a demand, that Elizabeth show up for today's pinochle game — all the women in the valley (ten or eleven) would be there, all with the winter *blahs* (you should hear her say the word), intending to do something about them. Becky said it didn't matter if anyone knew how to play or not, just to be there. I told her Elizabeth would be there.

I went home and told Elizabeth about it, and asked her not to appear too delighted, too enthralled by the winter weather, this snow and silence, that it wouldn't please the local ladies at all for her to be wintering so much better than they were. Plus, we're fresh back from two holiday weeks in Texas, Christmas and New Year's, and missed the five cold days of twenty below. We had driven south while the temperature was still falling, eight below, nine below, minus ten; we were gone a day ahead of the real cold. So who can say we wouldn't have the blahs too, if we had been here for that as well?

But I think not. That was a long time ago. It is a hoot to see the silence and isolation working its way on all the veterans, even the men, who are outside more often. Dick, Becky's husband, asked us the other day if we ever went bowling. He and Becky went into Libby every Sunday, and we were welcome to join them. These guys are about as social as corn dogs in the green-growing summer months and in the crisp excitement of fall, but biology is tightening the screw, twisting down on the tops of all our heads, and like puppets, we do what the strings tell us to keep from getting tangled.

The snow is slanting hard past the greenhouse windows now, blowing and swirling. Big flakes, a true blizzard.

Breitenstein, the old warlord, just drove by in his big red

truck, a holdover from the fifties like Breitenstein himself. I think that Breitenstein alone is immune to the turn of the globe, to the ways of civilization. He doesn't know it yet, but he's about to get a treat, a surprise.

My truck got stuck down in his larch bottoms last night when I was taking out my last load of the day. He'll be delighted, almost incoherent with incredulity at my pilgrim ways. With luck, I'll be able to hire him to pull my truck out. Breitenstein might, though, try to contain his excitement, play it for a day or two, and just let the truck sit there, waiting for me to come ask for help; or he might simply wait to see what I do, see if I wait until the spring thaw. If he makes that mistake, leaving my truck there, playing cat-and-mouse with his joy, then I'll go down this evening, jack it up, chain up, and drive out, sorely disappointing him.

But it *is* winter, and there's boredom in the valley. I suspect he'll chain up to it with his big truck, root it from the ditch, and drag it out from its spun-ice tomb all the way up the hill, leaving it in my front yard so he can crow about it for years — and he will have earned that right. I was down in the woods after dark, the city workaholic, trying for one more load, when I could have had all of today in which to work, and tomorrow, and beyond, if I had not gotten stuck.

Before I got stuck yesterday, I was cutting wood in the late afternoon, near dusk, when Breitenstein drove up. It was the first time I'd seen him since getting back from Texas, and he seemed irritated about something. I had bought a new chain for the saw and it was cutting straight again, and I had a big stack of split wood piled next to the truck and was making headway on my logs. I thought at first that he was annoyed that he couldn't find anything wrong with my operation, but I soon found out that wasn't the case.

Earlier in the winter, when he sold me the logs, he'd talked about building a deck. I was over at his house, listening to him go on about his deck, and not really understanding why he had such plans for it, such fervor, or even why he was telling me. It seemed out of place for Breitenstein to be bragging to me, or to anyone, about home improvements, and I must have nodded dumbly, said something like "A deck will be nice," or maybe said nothing at all.

Yesterday, however, I learned what a deck is, and where he planned to put it.

It's nothing more than a load of sawlogs for a log truck to pick up — like a deck of cards. I suppose, ready to be cut; a deck of timber, likewise — and he wanted to put it where my logs were, where they had been for over three weeks. He was amazed and a little angry, and not only because I was taking my sweet time getting my larch cut and out of there. I'd planned to take all winter with it, just for a hobby, a chance to go out and burn up a tank of chain saw gas once or twice a week, just for exercise, just to be in the woods; I'd gotten my wood in, forty cords now, with really no room for more; surely he, if anyone, with his daily putterings in the woods, no matter what the weather, could have understood that. But what he seemed most angered about was that I didn't know what a deck was.

I didn't try to apologize for my ignorance, though, or cover it up, and he said, "I guess you've never been in the woods before," and by "woods" he meant logging. I said, "Hell, I'm just learning how a chain saw works," and that seemed to take a little of the good red wind out of his lungs, my agreeing with him rather than arguing. Then Elizabeth, who had been skiing on the logging road, came sliding up and we got to talking about cross-country skiing, which I was surprised to find he'd never done, though he snowshoed everywhere. This started him off on a tirade about snowmobilers using the south fork road, traveling up and down it when it was against the law to use snowmobiles on a road that was still passable to trucks — and I thought how hard it must be for him, to see things done wrong in the valley, when for so long he was the only one living here, which was full guarantee that it would be done right, and done right the first time.

At any rate, he wants the logs out in a week. It turns out he wasn't planning to put the deck in until next week, and most of his anger had been the three weeks of fuming, *thinking* I was going to let the logs sit there all winter (which I was). I told him he should have come and told me he wanted them out right away, instead of sitting back there getting angry, but he didn't seem able to grasp this concept, and once again I can't say that I blame him.

You don't ask for anything in this country. Not for the time of day, not for help building your house, not for a tow when you're stuck. This is not to say that these things won't be offered, but the basics of communication — words — have never evolved out here, not yet, with the distances so great and the people so few, and I hope they never will. Actions are the way to communicate, not words.

It's simple. If someone *wants* to tell you what time it is, or *wants* to pull your truck out, he will. And if he doesn't, it's because he doesn't want to, and so it's pointless to ask. A country of mutes, but also of doers, not sayers.

And of paradoxes. Blowhards can come through and make a killing, can swindle the locals beyond belief. Everyone in the bar the other night was talking about Jesse Jackson, about what a great speaker he was, and how they were going to vote for him — not because of any great presidential qualifications or experience, but because he was such a great communicator, a great speaker.

I wanted to say, We did that last time. But I thought better of it and sipped my beer and looked around. It was warm. Jim and Tammy Bakker could be president and it wouldn't matter: they could never get to me in the Dirty Shame, in the Yaak Valley, in winter. I'm safe.

JANUARY 10

You notice how the days are getting longer again, how you've gained, in less than a month, an extra half hour in the evening alone. Your mind starts to wonder, daydreaming, and you begin to consider other seasons.

JANUARY 11

The woodpile has shrunk. Days grow longer, brighter, and the sun pours through slats in the shed, throwing tiger stripes against the empty walls. Cold air moves through the woodshed now that the wood is mostly gone. In the summer it will be a cool place. In the summer it'll be a place I don't visit at all.

Artie is behind the bar, drying glasses with a towel. I ask him what's the coldest he's ever been, hoping to hear a story of heroic, mind-numbing cold — crawling down Mount Henry with a broken leg in a blizzard, or crashing a snowmobile through thin lake ice, through spring-thaw ice, a long way from home — but he only looks annoyed.

"Shit, man," he says, looking out at the river, "ten below, forty below, it's all the same." He thinks I'm talking about mathematics, temperatures — numbers. "Cold's cold," he says, seeming somehow disappointed in me.

He looks out at the river, the bright sunlight shining on it, the ice breaking up, the river flowing again. It's like pain,

I think: you forget it soon, almost immediately. You try to remember it, but it's gone so quickly. The temperature's hovering right around twenty degrees in the sun. Shirtsleeve weather. You can almost feel the sun if you concentrate.

There's a point where you can give up on winter — when temptation can enter your soul, prying its way in like cold air through the cracks in your cabin — around January sixteenth or so, and this can make you realize that February's coming, and beyond February, March. For all practical purposes, save for a few blizzards, March will be it. April will come rushing on and winter will be over.

See, I don't yet realize that March will be the hardest month. Early February's the coldest, and often the snowiest, but March, strange, silent March, will be the hardest.

The danger in yielding to thoughts of spring — green grass, hikes, bare feet, lakes, fly-fishing, rivers, and sun, *hot* sun — is that once these thoughts enter your mind, you can't get them out.

Love the winter. Don't betray it. Be loyal.

When the spring gets here, love it, too — and then the summer.

But be loyal to the winter, all the way through — *all the way,* and with sincerity — or you'll find yourself high and dry, longing for a spring that's a long way off, and winter will have abandoned you, and in her place you'll have cabin fever, the worst.

The colder it gets, the more you've got to love it.

I think about all the couples up here: Mike and Julie, Bear and Lisa, Della and Fraecke, Robin and Jimmy, Dave and Suzie, Tom and Nancy, Dick and Rena, Dick and Becky. All of them are on their second or third marriages — Elizabeth and I remain the only ones who've never been married — and I think of how no one's ever been divorced once they get up here. These are not super-human people, superhuman relationships — the 100 per-cent prior-divorce rate speaks to their wildness, to their fear of fences, borders, control — but no one that I know of, since moving to the Yaak, has ever been divorced.

Doubtless, in the history of the world, there are couples who've divorced when they found *out* their mate wanted to move up here.

Learn to love the cold, the winter. If you love the country, the landscape — if you *really* love the country — then you may find yourself able to love it in winter most of all.

JANUARY 12

Nancy Orr — Tom's wife, the woman who makes the dream hoops — tans hides and sews vests and baby mocca-sins. She does the beadwork at the desk by her window looking out at the Yaak River. There's always a pot of chili or stew simmering on the wood stove in their cabin, which has antlers and hides hanging everywhere on the walls. Nancy is a pillar of stability in the valley, the acknowledged sage, the calm thinker.

One time Elizabeth and I were setting off on a trip — I can't remember where to; we were going to be gone for three or four days. When we got down to Libby — no easy task in the storm we'd come through — we couldn't re-member if we'd turned the generator off when we left the cabin. We'd had it running right beforehand, as I'd been doing some typing and Elizabeth had been ironing some scarves. But in a rural version of "Did you remember to take the clothes out of the washer?" we couldn't remember, and the more we thought about it, the more it seemed that we had not.

It would have taken us at least four hours to drive up there and check and then come down, just to get back to our starting point.

"Let's call Nancy," Elizabeth said, after a moment's thought. I hesitated, because Nancy and Tom live up in the town of Yaak, a little past the mercantile and the pay phones — that's how they got a phone in their cabin. I hesitated, too, because it's hard for me to ask anyone for anything, whether up here or in the city. Not that they wouldn't do it; it's just that I hate to admit that I messed up, especially up here, especially being new. I thought again of four useless hours, creeping through the snow, and nodded and said okay. Miraculously, Nancy was in when Elizabeth called, and they got it straightened out. Nancy said she'd go right down and check, and Elizabeth could call back in a half hour and see whether it had been left on or not.

We had a cup of coffee and some pie in Libby while we watched the clock and waited.

"She told me something funny," Elizabeth said. "She said I was the third person this year who'd called from out of town and asked her to go check their generator."

I smiled and felt warm, less foolish, but also a little guiltier. There are four or five places in the town of Yaak that, because of the pay phones, have their own phone lines: the mercantile, the saloon, and a couple of other cabins. I felt bad that it was always Nancy who was getting called, but guessed that was what she deserved for being responsible and dependable.

Nancy and Tom use the bones of deer and elk, cut in cross section, to make shirt buttons. Tom hunts with a homemade bow and arrows made of cedar, and always shoots a big buck. One day back in the fall I saw Tom, who used to be a champion bronc rider, walking down the road in his moccasins, which Nancy had made for him, smoking a pipe (one he'd carved), and holding a big red-tailed hawk on his arm while his dog, Nuthin', a German shorthair, ran along beside him, sniffing for grouse — the dog would flush out the birds, I presumed, and the hawk would catch them. I drove past and waved, and watched Tom walking and smoking that pipe — walking into another century, and whether into the past or the future I couldn't tell, but I'm glad he's in this valley.

When we called Nancy back — she caught the phone just as she returned to the cabin, out of breath — she said that we had indeed turned the generator off.

"It's snowing hard here," Elizabeth said.

"It's snowing hard here too," Nancy said.

The world can be so safe in winter. I was homesick for Yaak, and we had barely left.

JANUARY 13

If you look out at the snow, and beyond it, trying to see through it to the woods on the other side of the meadow, it seems to come down fast, and your life, if you let it trick you that way, seems to be just as hurried and frantic. But if you remember to look at the snow like a child, or a Texan — gazing up, trying to see where it originates — then the slowness into which it falls, the paralysis of its journey, will drop you immediately into a lower, slower state, one where you're sure to live twice as long, and see twice as many things, and be two times as happy at the end. Snow's more wonderful than rain, than anything.

Ravens call out as they fly through snow. They're surprised by it, I think, it starts up so quickly — one second a gunmetal sky, and the next all the snow emptying out of that gray color, tumbling down. That touches new corners of my brain, things never before seen or even imagined: the sight of a raven flying low through a heavy snowstorm, his coal-black, ragged shape winging through the white, the world trying to turn him upside down, trying to bury him, but his force, his speed, cutting through all that snow, all that white, and headed for the dark woods, for safety. For a few beautiful moments there's nothing in my mind but black, raven, and white. My mind never clearer, never emptier.

There's a strange thing about myself that I can't explain. When I walk in the woods, if it's not snowing, I feel like exactly what I am — a man, alone, walking in the woods. But if snow's falling — if it's snowing heavily, with that underwater hush everywhere, that cotton-stuffed-in-one's-ears silence of soft snow falling — I feel like an animal.

I've seen flakes as big as my fist, and monstrous, wet, stuck-together flakes as big as wadded-up sheets of newspaper, falling among the myriad of other, smaller flakes — plummeting madly down, tumbling like planes crashing, but landing silently, or nearly so. I'm describing the onset of a blizzard, which is what it's beginning to do right outside the greenhouse window and over my home, my cabin, and Elizabeth one hundred yards down the trail.

I can still see the outline of the house, and in it all that is my life, the shape of it, through the heavy falling snow, but just barely.

Take nothing for granted.

JANUARY 14

You go long stretches without seeing any game at all, not even deer, much less moose or elk, coyote or lion (bears are hibernating, rousing briefly this time of year to give birth to pea-size cubs before going back to sleep, lactating in their sleep). It's not a good feeling — where have all the animals gone? You don't even see their tracks in the snowy woods, except for the ubiquitous, ever circling hares; silly, wandering tracks. Then suddenly you'll be blessed again: you'll see cow moose, sometimes with calves, or just confused, gangly calves fresh out on their own; large herds of deer standing back in the trees; and herds of elk in the trees beyond the house, having been chased lower still by the continuing high snows. The dogs growl at them at night, and the elk (watching your cabin lights as if they were lonely) will trot away like horses, hooves striking rocks and logs, grunting, moving on as a herd.

Last night Elizabeth's best friend, Julie, with whom she paints and plays pinochle, and with whom she plans to garden in the spring, told us that she and her husband, Mike, are expecting a child. We were at the saloon when they told us, and, like that raven flying through the snow, it was an eerie feeling. We congratulated them and knew they'd been talking for ages about wanting a baby — Julie had been, anyway — but still, it came as a surprise.

August, said Julie, the month when the women in the valley pick huckleberries.

Elizabeth and I were quiet the rest of the evening, freighted with a new, terrible sadness. We felt so dissociated from everything. Plaster smiles, weak smiles, poor conversations with whomever we sat next to. Everyone else so terribly gay, and so much noise and light.

Change always hits us hard. We're delighted for Mike and Julie. It's just that the four of us were so close. It's simple childish jealousy, I think — though there's no true jealousy, just childish sadness. Or adult sadness. Or some damn thing. God, we were sad, without knowing why — and still not knowing why.

Is it that all four of us, Mike and Julie and Elizabeth and I, were up here to hide out, to hole up, to build a fortress against the rest of the world? And that now, suddenly, over the space of one night, one sentence uttered — "We're going to have a baby" — they have gone over to the other side? That they're *of* the world once more, of necessity?

Julie said that the night she told Mike about her pregnancy, they were in bed holding hands and a moose came right up to the bedroom window, put his nose against the glass, and fogged it up. They shone a flashlight on him and screamed, but then laughed.

It was as if he'd come to see the baby already, Julie said.

I try to pretend that the days go by slowly, and sometimes they do. They hang back, the way we want them to.

There's a football game on at the Dirty Shame today, a playoff game. We'll go up and watch the second half of it, and drink a beer or two.

JANUARY 15

My mother's birthday. It's snowing hard. I drove up to the pay phone to wish her happy birthday, but she and my father had gone out for dinner. I drove back to the cabin through that falling snow, driving carefully, and thought about the last time I'd seen her — Christmas — and looked ahead to the next time I'd see her.

I finally got the truck out of the larch bottoms. Breitenstein never got a chance to see it: there were no tracks leading into the second turn-off, the one I use. When he does his cutting, he goes into the woods at the first turn-off, so he never suspected a thing. My virtue and integrity as a woodsman, such as they are, are intact. I am not ruined yet.

Elizabeth said the pinochle ladies were fun, but they smoked a lot. Elizabeth said the ladies told her that she caught on quickly; she even won a couple of hands. We had both expected mounds and mounds of gossip, and I'd asked Elizabeth to take notes, but was disappointed. She said mostly they just played pinochle, and were damn serious about it.

JANUARY 16

A freak January thaw. The rain is the worst: not true rain but a fine falling mist, looking like snow, sometimes even flurrying. In our hearts, we know it is so close to rain that it doesn't matter what it looks like, and when it hits the roads, our arms, anything, it melts, leaves water, then later forms into ice. It is a slow-falling rain that would be beautiful in the late fall, with a fire going and a book, but the few people staying in the Yaak this winter are crazed, want to get out, and even a blizzard would be better than this, the hideous thaw.

I went up to the pay phone to make some business calls last night. Dick was drinking beer and eating smoked salmon and sturgeon. He kept giving me fist-size bricks of it, telling me how good it was, especially with cold beer. The fish was dry and very salty, and indeed it was good with the beer. I'd only stepped in to buy another writing notebook and a cup of yogurt, but Dick corralled me and began talking about all manner of things, offering me the fish, beer, and wisdom. Though I may have taken advantage of him, and he may not remember it in the morning, I did manage to wangle an offer of sorts from him to take me ice fishing over this special hole he knows about in the Yaak River. I also learned the name of his favorite lake up in the mountains, Hoskins Lake, and how it's good in the winter too, but best in the fall.

Victor, Dick's son, whom I'd not seen in a month, came into the store. He was doing his usual chores: carrying out garbage, plowing snow and ice away from the gas pumps, flattening — eternally flattening, it seems — aluminum cans. Sometimes he drives back and forth over them with big tractor wheels, but occasionally, when he is feeling mellow, he smashes them with a stump that's got two spikes driven into it, for handles, which he uses like a jackhammer, demolishing them one by one with a satisfactory tinny sound, the gold and green and red cans glittering, smashed flat as wafers, a penny a can — his trip to Hawaii, perhaps, if he can get enough of them.

I stood in the mist for a while with Victor, who seemed very pleased to see me. He slugged me on the shoulder and asked how I was doing, a greeting I imagined he usually reserved only for his high school buddies, and it made me feel young. He was wearing a new black-and-red-plaid lumberjack's overshirt, and I felt I had to give him something back, so I complimented him on that. I helped him crush cans as we talked, crunching them with my boots, then kicking them soccer-style into the big, growing mound.

Victor learned to downhill ski only last year, and had plans to leave on the twenty-third for a place called Mount

Schweitzer, north of Bonners Ferry, and was already gearing up for that. It seemed so paradoxical that here, in the heart of snow, he had never skied before. I was used to people in the South making $5,000 pilgrimages, using all of their vacation time and money to fly up to this country and roll in the snow for a while, and I had always imagined that the lucky ones who lived up here merely woke up and decided, on any day, at the last minute, to head up the nearest hill and make a few runs, the way a city dweller might decide, on the spur of the moment, to go for a jog in the park.

I've noticed that the young guys in the weight room down in Libby are the same way, though. Like high school girls, they'll talk about an upcoming trip that is two, three weeks away, as if it were the senior prom or something, rather than just getting out and *doing* it. When my friend Kirby and I were in high school, we'd get that same three-week restlessness, but weren't disciplined enough to wait: we'd have to do something right away. We'd pool our money, drive out to the Houston airport, and catch a ten P.M. flight to New Orleans for $59 round-trip, and we'd eat French doughnuts when we got there, wander the streets, look in on the female impersonators, play pool, go pay homage to the river, ride the streetcars all over, then catch the four A.M. flight back home before daylight. We'd sleep until nine or ten, with no one the wiser, no one the poorer, and the itch, the unhealthy, bottled-up, crazy-making itch, would be out of our system for a while.

The best thing Victor told me last night about the country was how, when it's forty below, you should put cardboard in front of your car's radiator to hold the heat in and keep the cold air from blowing in. He drives his truck eighteen miles down the Yaak River road, where he catches the school bus. I remarked how horrible that must be, waiting for the bus in forty-below weather, but he said no, that the bus was usually already there.

I picture the bus driver waiting an hour, wind shaking and rocking the bus, engine idling, reading a paperback patiently as he waits for the high school students from the Yaak — all five of them — to come straggling in. He'd be running the heater, eating an apple maybe, and in no great hurry, with snow swirling down outside.

A slow country. A country from a long time ago. It's easier to learn certain things when you're watching them occur in slow motion.

JANUARY 17

A bright blue day, fresh snow high on the mountains, on the trees, the far forests frosted with it, and crystals sparkling in the air. I believe the old Jim Bridger legend about the time he wintered in Yellowstone country. He'd later gone back east and told city people about words coming out frozen when the trappers tried to speak; about how they couldn't hear each other because the words froze the second they were out of their mouths — and of how they'd have to gather the frozen words up and take them back to the campfire that evening to thaw them out, to hear what had been said during the day, stringing the words back into sentences. I can picture it being that cold.

Ice flakes sparkle in the sun on the sunny blue cold days up here; no snow, not a cloud in the sky, but little flecks of frozen moisture, tiny ice crystals like glitter, flashing and sparkling. Breitenstein had said that when it's really cold, these minute crystals collide in the breeze and make a faint tinkling sound, like chimes, like glass clinking, a magical sound.

I can see a winter so cold, a day in the woods so distant and away from warmth, that the moisture in the breath of one's words freezes, making that tinkling sound as the breath comes from one's mouth. I can see a winter so lonely, a season of silence so long, that perhaps the habits of speech are forgotten, and a trapper might *think* he was speaking, as one imagines oneself to be speaking in a dream, and nothing but the tinkling of glass would come out, another language being spoken, a language of winter, of determination, of purity, frozen.

Bridger's stories were tall tales, but the tallest are usually the truest. No one believed the tales about how hell had opened up and steam came rushing through holes in the ground; how he took steam baths in the dead of winter, in the one place that was safe from Indians, because of their fear of the area, and where he could sit back in the bath and relax in the middle of the wilderness, naked in a pocket of warmth. But these stories were true too.

The other night I saw a moose come running across the field, dragging someone's barbed-wire fence with him. I watched as he galloped in long, easy strides across our pasture. He took out Holger's fence to add to his collection, crossed the road, and then took out Holger's other fence, dragging perhaps fifty yards of fence and posts with him. It was dusk and he kept going, seemingly not running *from* anything but as if he were simply in a hurry to *get* somewhere. He disappeared into the woods, and the next day I tracked him.

I followed his trail for a mile, a swath through the woods, young saplings and ferns plowed up and razored level as if in a clearcut, a tunnel of light through the trees; I never found any fence, never caught up with him. This reminded me somehow of the great children's story of Paul Bunyan and his blue ox, Babe. I think all of the old stories are true, that there was a John Henry, and a Hercules.

Breitenstein drove by at about ten o'clock this morning. I was inside the greenhouse, struggling over my second page of fiction and into my third cup of coffee, my fifth log in the stove. I'd been dreading Breitenstein's passage, unable to get it entirely out of mind, and from a long way off my ears picked up the ragged-smooth sputtering of his old red truck coming up from the valley. Guilt washed in like a cabin door left open on a cold night — there would be no more good, unfettered writing this morning. I wondered if I should at least continue to try, but gave up.

Last night I finished cutting, splitting, and hauling out the last of the larch I'd bought from Breitenstein. He's a fanatic for tight fences; he stretches the gap into his larch bottoms so tight that it's impossible to open it by hand. He has a lever-stick wired to the post, so you have to pry the loop off the gap with that, and even then you must lean into

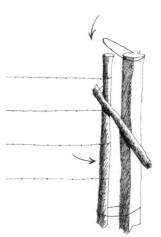

it with all your back, grunting and cursing, just to unwedge the little wire loop that joins the gap to the rest of the fence. When you finally manage to lever the wire loop up, the gap eases off with a sigh of relief, and you lay it down in the snow and drive on in to cut your wood. Buried in snow, the barbed wire contracts stubbornly. Days later, when you're through, you've got an inch or two of shrinkage, and there's no man born of woman who can close the fence back up, not with the lever, not with his back, and not even with a longer pry bar: the fence posts begin to creak and splinter, you're pulling so hard, but that cold barbed wire will not stretch.

In Texas, going from ranch to ranch, where all the land was privately owned and had some livestock on it, I was taught never to leave a gap open: never, not for any length of time, because as sure as anything, an animal would get out. In writing I've been taught the same thing: always close your gates — unless, of course, you *want* something to get out, escape into the next chapter, perhaps, or even into the night, never to be seen or heard from again. The things that count, however — the goods, the story, the livestock — those you always keep the gap closed on, or at least close it when you are through with them.

But last night, with the fence shrinkage, I couldn't close the gap in Breitenstein's fence. I couldn't picture him being any stronger than I was, couldn't picture *anyone* being so strong. Yet I knew how much it would anger me to have someone come onto my property and then leave with the gap wide open, and I'm not even a fanatic about it, the way Breitenstein is. I prefer the gaps to be loose and a little floppy — easy to open, easy to close, but tight enough to keep the animals from realizing there's a weak point in the fence.

What the fence looked like before I opened it

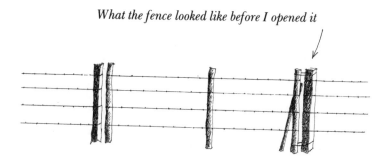

Breitenstein doesn't even own any cattle. All he has on the other side of that fence is trees. But a gap is a gap, so I spent most of the night (stars, but no moon; the old snow freezing again) making hard crunches as I dug in, leaned against, begged, cursed, slipped and fell, got back up again, working on that damn piano-wire gap. I couldn't get it, so I gave up and just rigged it, built a larger, makeshift loop with which to catch the gap's last post.

What it looked like when I was through

But I didn't sleep well, knowing I had done the job poorly. I kept imagining what my father would have done back in Texas, and how he would have fixed it somehow, made it right, even if he'd had to stay out there until daylight — he might have built warming fires all along the fence, to cause the wire to expand again, loosen it back up. I imagined, too, how *his* father would have done the same, and his father's father, and I felt very guilty for being so critical of the decline in American standards, American values, American everything, if I was no better, if I was as much a part of that decline as anyone else — I had just *quit* on that fence gap, no two ways about it. I had given up and gone home and eaten supper, and tried to read for a little while, warm in my house. So this morning, when I was half-writing, half-listening for Breitenstein's old truck, I knew when I heard it that even though my first priority, always, was writing, and writing well, I could also feel a strange conflict, and I got to thinking about what my father would have done, and how right was right and wrong was wrong, and how even if by some wild chance he *hadn't* been able to close the gap the night before, he would have been down there with Breitenstein, asking if he could help, letting Breitenstein know that he hadn't just walked off the job, that he understood it was important and was going to see if the two of them together could get it . . .

And so even though the writing was more important than Breitenstein could ever be, more important than anyone could ever be — which is sad, but the way it has to be — I found myself closing down the wood stove, finishing my coffee, and leaving my notebook open to the page I was working on. I went out to the garage, started up the truck, and drove off down the road after Breitenstein, as if controlled from above.

I've read of other men my age experiencing this same thing — all men, it seems — one day becoming themselves, but also their fathers. Just like with legends, men build the base of themselves with parts of their fathers, with the basic truths — never leave gaps open — and then go on from there, of necessity, to alter other things, grow in new places, and become fathers themselves — but always growing out of the basic truths, a few simple things. And when you are twenty-five or thirty, and the world's becoming smaller, and things are starting to move, finally, like a train slowly leaving the station, picking up a little power, a little speed, that is when, at first, you're lifted by the bootstraps as you're going out the door, pretending you are him, doing the right thing, doing what he would do, until in the end, by rote, you learn it, and just in time.

When I caught up with Breitenstein, he was already at the gap, was already walking back from it, and I feared his wrath as surely as I would have feared my own. Leaving another man's gap down, whether he has any livestock or not — it's careless, it's wrong, it's thoughtless. But he was glad to see me.

He seemed unconcerned that I had not been able to close it properly. I was trying to explain, but he pulled a long yellow tool from the back of his truck, a piece of heavy metal that looked like a bumper jack with a ratchet handle on it — a fence stretcher — and showed me his secret, how he did it. I knew of fence stretchers, had seen them before, but did not understand that they could be used on gaps as well as for just stringing fence. Breitenstein seemed calm, benign, even for Breitenstein. I later realized he was pleased that I had at least bothered to rig the gap, however loosely, until later, and that I drove down then to see him.

The morning was cold and bright, and we talked for a

half hour. He said he wasn't going to cut the aspens in his meadow; he liked the way they looked. (Earlier in the winter he had talked about cutting them down for paper.) He pointed out good mountains to hike, told me about different kinds of snowshoes, about trap lines, V-6s, peavey hooks and cat hooks (logging tools), stubble discs and bad farmers. He was delighted that I knew a little about snowshoes and could speak a sentence or two about them, but quickly cut me off and went into a tirade about the hippies, dope fiends, and wood thieves whom he had encountered recently, all of them bad except for one who had helped him put a bridge back over the county road once when the old one washed out: "He's okay, he works for a living."

Then Breitenstein raged some more, as we stood in the road, about people who had owned the land before him, and it had done me almost as much good, to learn about peavey hooks and cat hooks, as to listen to Breitenstein curse and dismantle those people. It was as if Breitenstein were routing the lazy ones from a field, like bad cattle, and shutting the gap behind them. I wanted to tell him about my father, and about closing gaps in great novels, how the good ones were strung so tight they would stand forever. But Breitenstein went on and on, the friendliest he'd ever been with me — and then he ended our conversation quickly, as if electrified: "Whoop, I got to go to work!" And he added a most uncharacteristic "Have a nice day." I told him that I would, and thought of how my father had gotten me through another day, 2,500 miles from home, and of how eventually, soon perhaps, I would be floating — until then I was still sometimes a boy, only imitating a man — it might take years, but I was on the right track, and I could see up ahead where I was going.

JANUARY 18

McAffee, a hunter, said he drove past the house before daylight the other morning and saw two bull elk and a cow eating hay with the horses. Everyone at the bar became excited over this. So far, I've seen only their tracks. I'll have to get up even earlier tomorrow.

I got drunk on eleven beers at the Dirty Shame, watching Denver barely beat Cleveland (Kosar went fifteen-for-seventeen in the second half; unbelievable), and then ate two Dirty Burgers and a bowl of Shameful Fries. I went home and reread a novel (*The Bushwhacked Piano*) straight through, start to finish.

JANUARY 20

Set the alarm clock for seven, got up, went out to the greenhouse, and did a morning's work by candlelight. A new man. I have to go into town today and shop, do errands, sign papers. If only I could shed that other life, the going-into-town life, like a cicada, pulling free from a tightening, drying, constraining old shell, a molt. But an old one always seems to grow back.

A driving snowstorm, big flakes blowing past, crashing into the woods, swirling in the meadows. They are the currency of winter, and I am the richest man in the world.

JANUARY 24

Cold last night, eighteen below, and only two above now, at noon. We keep the fireplace stuffed with larch. How something can be so useful after death is amazing. We went downhill skiing yesterday over near Kalispell and had a blast: Elizabeth's first time.

It is almost February, the snowiest month.

JANUARY 31

There was a letter yesterday from Kirby, saying that already the baby was kicking.

Sticks of wood, burning up like days.

FEBRUARY 1

Twenty-five below, and I'm trying to write without gloves on next to the groaning, stretching stove. Runny nose as I thaw. At twenty below, the sticks of firewood feel supernaturally light. All the moisture has been freeze-dried out of them. The big logs in the woodshed, when I pull them down to carry them into the house, make clinking noises, as if they're being split. At twenty below, you could make a giant mobile to hang in the still air, 60- and 70- and 125-pound chunks of larch dangling from a rafter beam, and when the winds started, they would sound like glass, tinkling against one another.

That last sentence cost me. I have to write a few words, stop, and tuck my hands under my arms to warm them. I wrote that one all the way through. Got to stop now.

I sprinkled old coffee grounds all over the greenhouse beds of earth today. I'm looking forward to turning up the soil, growing things. We had some good carrots last fall when we moved in, before the first few freezes got too hard. This is good lettuce-growing country too — at least in the garden right outside the greenhouse. Doesn't ever get too hot.

FEBRUARY 3

Later in the day. Watching log trucks creep by every so often. Snow-caked, limb-stripped trees are stacked so neatly, like toys, on the trailers; the trucks move slowly up the hill, as if off to war. All the trees are so slender, so small, like matchsticks. I haven't seen a big tree on a single load all winter long. It's terrifying. I'd rather they were cutting young trees instead of old ones, but I know the score, and realize that the reason I'm not seeing any big logs taken out of the valley is that there aren't many left — not that they can get to easily, by existing roads, anyway.

But I still know where a few are, some that I like to sit beneath in the fall.

It's taken me a long time, but I'm finally learning.

Put stumps in back of your pickup to give weight and bite to the rear axle, to keep from fishtailing in deep snow. Tack cardboard in front of your truck's or car's radiator to keep the engine heat in.

These back-up parts for the generator, most necessary to have on hand: extra points, spark plugs, oil filter, fuel filter.

Battery charger and extra charged battery to carry out to ailing truck, like a replacement heart.

Snow chains.

Sleeping bag, ax, in all vehicles.

Gloves, flashlight.

Know how to shut the water off if the pipes freeze and then thaw and break.

Snow shovel in all trucks.

Don't set your emergency brake when parking in really cold weather — it could freeze that way, and you'd be stuck.

FEBRUARY 9

Dark and snowy morning. Writing by lantern light; wood stove snapping and creaking, expanding. Still shivering a little. Thought about summer a lot last night: about fly-fishing, about sitting out on the porch in the evenings drinking cuba libres and watching the mountains.

FEBRUARY 10

Driving back from Libby late at night, having stopped five miles up the Pipe Creek road (another thirty-five miles left to go) to put the snow chains on — deep, dry cake snow on the road, and snow floating down, trying to erase the road. It's late and we've been out to dinner, but now we're headed home, being drawn farther and farther back into the wilderness. When we start driving again (good music all around us — Joni Mitchell and then John Prine — the cold snowy night pressing in like the greatest friend), the road begins to narrow, weird-sculpted snow bluffs looming over either side of the mountain road, so that it's like driving through a tunnel. The road narrowing and narrowing between the baroque, towering drifts (so high above the truck that if they collapsed, in an avalanche, we'd be swallowed, and the road truly *would* disappear) until there's barely room for us to fit. The snow walls are scraping either side of the truck's doors, and the wind's swirling in a fast fine crazy mist, blowing sparkles of spray across the beams of our headlights, chains blasting the deep, loose snow on what used to be the road into shifting white dunes — and all this for a steak.

Wiser souls would've stayed in town, checked into a hotel room and played pinochle, lay in bed and watched, in slovenly torpor, the hideous television, and would have had a fine time doing so — and we considered it. But the snowy road, and the night beyond the cabin, lured us: home.

"Are you watching this?" I ask. Elizabeth nods. We're rolling over the windy summit, barely able to see what's left of the road — two-foot-high dunes are being reassembled in the great wind, and no other tire tracks out, not by a long shot. When we cross over the summit and begin the descent, the wind changes direction, so that it's blasting all that headlight snow-glitter straight down the tiny tunnel of a road, sparkling frost and ice crystals rushing alongside us, and we're caught up in the race of it, with that wind at our back. There's nowhere to go but along with it, down the summit, down the wind tunnel, racing along with the eerie, rushing frost ice, the spooky ghostliness of it.

"It's like *we're* the wind," Elizabeth says.

It's after midnight. We hum along, coasting in four-wheel drive down the mountain's face, with the high cliff on one side and the gorge on the other. The chains are flopping regularly, a hypnotic sound, giving our tires good bite into the shifting snow, and it is a lovely sound, jingling like Christmas bells, like reindeer sleigh bells. We feel like kids: it's night and the world is ours.

It's a strange business, trying to tell you what this valley's like. I've got to tell you about Don and Kathy Terlinde, who are our second-nearest neighbors and live about a mile away. They're shy, private people. I've seen them maybe three times this year: once at Thanksgiving and a few other times.

Don is toothless, wizened, and has a crew cut. He plays the accordion every Thanksgiving, plays "Popeye the Sailor Man," and looks, in fact, like Popeye, with his eternal, wind-blasted, tanned grin. I went over to his cabin one time in the fall — drove down the road to its end and then walked in the rest of the way, unannounced — to see if he'd seen a big red dog, an Irish setter I was babysitting for a friend. All during the day, I'd been hearing Don sawing and hammering something down through the trees, toward his cabin, and I got to thinking that maybe Isabelle, the beautiful red runaway bitch (she was driving me loopy!) had gone over there to check out the noise.

I hadn't heard any hammering or sawing for an hour or so, though, and when I walked up, I saw why. Don was lying on the side of a hill in some pine needles, in the afternoon autumn sun with his hands behind his head, looking at his cabin. He was building a porch, and there were boards, logs, sacks of concrete, and sawdust everywhere. He was just sitting there grinning, not napping but grinning, looking at the project he was working on. He didn't hear me coming, so I surprised him, walked right up on him with the wind blowing the larch needles through the air.

He started to sit up quickly when he saw me, and then grinned wider and lay back down. "Just studying it," he said, nodding toward the porch.

FEBRUARY 12

This sound: dripping. Running water coming off all the roofs, down all the downspouts, trickling across the paths and game trails in rivulets. Windy, with gusts and swirls from every direction, but mostly from the south and west, and purple skies, yellow weeds sticking up through the snow and blowing back and forth in the warm winds. A chinook: forty-eight, almost fifty degrees above. I am in the greenhouse, listening to the light rain on the roof, and today it is more like Mississippi than Mississippi.

Spring is suddenly here, like the new kid in town who merely shows up one day in the middle of the semester, who talks funny and can play baseball better than anyone else. Winter is being routed in a single day.

I understand that it is early February, that this is only a false spring; I understand it but do not believe it.

Like a stock speculator dealing in futures or paper gains, I am thinking ahead to all of the things I can do with spring.

One almost always thinks of winter as being the dominant, most severe season, the one that constricts all — but I never had any idea that spring could be so strong too.

The snow is turning to water and *running* down the hill. Great slabs and cakes of ice are breaking off and sliding down the steep roofs of the cabins, the barns, the greenhouse. The wind is dancing around in the yard and howling, spinning warmth everywhere. I can see things that were buried in the snow, things I'd forgotten about long ago, back in the fall.

If this is the false spring, I want to see what the real one is like.

FEBRUARY 13

I just finished the novel, the first draft. I need to remember that the first draft is only a false one.

Thirty-two degrees today. I feel a little lost. Should I build a fire and risk getting too hot?

I don't have any strong ideas leaping to get out. What should I next work on while the draft of the novel rests? It was easy during the novel — I knew who was playing, what to work on every day. Now I am alone.

FEBRUARY 14

They say that when the winter ends, it does so overnight, with a chinook. They say a summer-warm, purple rain comes roaring up the valley from the south, bending the trees in the night, and that all the snow on roofs, on the tops of cars and trucks, and in the trees melts: you wake up and it's gone — all is different, changed, lost.

FEBRUARY 15

Walking on snowshoes down a deer trail in the woods up behind the house. I can feel spring's coming, and I'm almost frantic, almost frightened, to realize winter's going. Sometimes I want summer and green grass to come, but sometimes I feel like I've committed my life to winter, moving up here, and what's more, I've fallen in love with it, and have gotten used to it, and can't picture there not being snow under my feet.

I feel as if I've gotten dependent on it, the way we always do with a thing we love. The natural terror of that thing not always being around.

Smoke is coming from our cabin below. Elizabeth's built a fire. She's ready for spring, and then summer. I need to be more like her, that way.

I need to let this snow melt beneath my feet (as if I have anything to do with it). I need to be able to just let it go.

Whenever anyone asks where I live, and I tell them, they say, "Do you live there alone?" I tell them, "No, my girlfriend lives there with me."

"Oh, well, that's different," they say, relieved. "That's good."

And they're right. It's still isolation, and still we're alone. But it is different. It's, well, perfect. Fairy tale.

MARCH 3

Breakup. Not Elizabeth and me but a bond nearly as strong: the bond between seasons, winter and spring, the bond beginning to separate, loosen, buckle and fold. Frost heaves in the barns, on the roads — the earth is stretching, coming back to life.

Elizabeth saw a bluebird flying through the woods the other day, the first "real" bird, rather than the silent and angry winter warriors who've stayed on — the great gray owl, ravens, eagles, and thermal-king grouse. The first *real* bird, one whose sole purpose is to sing and splash color across the land, to spread wild beauty.

I, too, feel myself beginning to buckle, to stretch.

The roads are thawing, losing the frost that has been locked into the ground's pores. The ice is turning to water, expanding, lifting the roads and the fields into waves, into a soft sea of mud. No logging trucks can negotiate the roads.

For roughly six weeks, after the logging roads go from their hard-frozen concretia to deep red mud, the logging trucks won't run — there's a ban on vehicles weighing more than ten thousand pounds — and so the trees will remain standing, and taxpayers will get temporary relief, too, from subsidized logging of their lands.

My days will be silent. No cars, no trucks, no chain saws, no nothing. I'm resting. The woods are resting. Birds are coming back.

It's still winter. I need to stay loyal to it. But I dream of grass.

I want to *eat* grass, want to walk in it barefoot, roll on my back, shirtless, in it. I want to rip it up and roll it into loose wads and throw it, have grass fights with Elizabeth, the way, five months ago, we had our first snowball fight.

It is hard to fit nature. A seam is coming up. I am going to have to step from winter into spring. I must do it gracefully.

MARCH 4

I ran a hoe through the stiff dry bare earth of the greenhouse this morning. I say this with almost a shudder. It felt like as strong an act of infidelity as can be imagined. I have got to hold myself together.

The motion of it: pulling the sharp hoe through the earth, the earth furrowing and rising, a trail appearing, and then chopping at the earth, and the earth turning up, loose and willing, soft, under the surface, piling up, ready to be planted.

I just got up from the desk, went in there, and did it. Now I feel guilty, guilty as hell — not at having done it, but at how good it felt.

I felt it all the way up my arms, all the way up into my shoulders, and even into my chest — and I still feel it now. It makes me feel weak. It feels wonderful.

I'm still in the greenhouse, back at the desk, by the stove, and a fly is buzzing at the window.

MARCH 7

Ten-degree nights, fifty-degree days. The hot sun is warm on my face. The snow is melting, water pouring off cabin roofs and barns in sheets, showers of water in the bright sun, the sound of running water everywhere — blue skies, white snowfields.

When I drove down to Libby yesterday, and passed the lumber mill, teenagers were parked out on the gravel island in the river, having driven across the shallows to get there. They were changing the oil in their cars, washing and waxing them. The blue water ran fast and away, with the white gravel of the islands shining in the sun. I looked away and thought of how, forty miles away in the woods, Yaak would be waiting for me.

There are two worlds for me — and for anybody, I think — and I do better in one than in the other. I used to be able to exist in both, but as I pay more and more attention to the one world, the world of the woods and of this valley, I find myself, each day, less and less able to operate in the other world.

Winter's a time for dreams. Do bears and other animals dream as they hibernate?

When I got back from Libby, I went up into the woods behind the house to relax, and I imagined I could feel the animals, and the woods themselves, stirring. The snow was melting, trickling — I could, and can, hear it everywhere. I don't belong in a city, and I'm not even sure I belong in the spring and summer.

The fall, maybe.

I've learned things over the course of this winter, this dream season, and have forgotten other things, old things that I'll no longer be needing.

MARCH 14

My parents came out for a couple of days, to see this new place Elizabeth and I have moved to. No one in our family has ever lived outside the South. We sat on the porch and had a couple of drinks, and looked out at the meadow and the mountains. I felt like I'd waited all my life to peel off my city ways, city life, and get into the woods — molting, like an insect or a snake.

Later that day my father and I walked across the meadow and went fishing in the south fork. It wasn't river-fishing season yet, but you could catch and release the fish, which was all we were interested in anyway. We walked through an old abandoned homestead — the great barn falling down, larch logs cracked and sagging in the middle — and crouched on a splintered, rotting bridge that spanned the little creek. Willows had grown up all along the banks. We cast our flies into the current, trying to catch little brook trout. There was still snow in the fields, but there were patches of earth too, and the sun was warm, almost hot.

Mayflies rose off the creek in small clouds. We caught several of the hungry little trout, and tossed them back into the cold water, and watched them dart away.

A bright-eyed weasel, with the first of his summer brown beginning to show across his back and legs, came out of the willows and ran fearlessly down the old bridge toward us. Then he sat down not ten feet away from us and watched as we threw the fish back.

"You've changed," my father said, not uncomfortably, as he mended his line. Across the meadow, against the trees, we could see the cabin. It seemed odd not to see smoke rising from the chimney.

"No I haven't," I said, just as comfortably, still casting to the little fish. We were catching them almost at will. I suppose I was pretending that I had always realized what I needed — deep, dark woods, and a quietness, a slowness — and that I hadn't been floundering for thirty years trying to figure this out, trying to get along in cities, trying to move fast.

He was right, though. I have changed. I can take apart a generator and put it back together. I can file a saw.

My heart has changed too. I'm in less of a hurry.

We fished a little longer, and I found myself thinking about the great gray owl, the way he caught mice last fall. I thought about those kids changing their oil down on the river, rinsing their soap suds into the clean river.

Right before my father and I left the little bridge, I tossed the last young trout I caught to the weasel, who pounced on it and then ran off into the bushes, his eyes bright. I could hear bones crunching.

My father and I walked back toward the cabin in the bright sun, stepping around the puddles and mud and patches of snow. It hadn't bothered me to hear those bones cracking, to hear the sound of the weasel eating the fish. I don't know why I did that, other than that the weasel looked hungry, after the long winter.

We could hear the creek gurgling as we walked. It sounded as if winter would never be here again.

Six or seven months can seem like an eternity. I suspect that, in the next two seasons to come, I may even forget entirely what winter is like, but then, like an animal — I hope — I'll remember.

Spring and summer will be okay, even though there'll be more people in the valley, more people everywhere.

If you go slowly enough, six or seven months *is* an eternity, if you let it be — if you forget old things, and learn new ones. Even a week can last forever.

Winter covers some things and reveals others. I admire the weasels, the rabbits, and the other wild creatures that can change with the seasons, that can change almost overnight. It's taken me a long time to change completely — thirty years — but now that I've changed, I don't have an interest in turning back.

I won't be leaving this valley.